Sir William Paton is Emeritus Professor of Pharmacology at the University of Oxford and an Emeritus Fellow of Balliol College.

His career in pharmacology included the discovery of the hypotensive drug hexamethonium, which opened the door to the treatment of malignant hypertension. Other work has involved the curares (being curarized himself), allergies, diving physiology, anaesthetics, and the pharmacology of morphine and cannabis.

William Paton has served on the Medical Research Council, the Council of the Royal Society, and the Wellcome Trust. As well as chairing the Institute of Animal Technicians and the British Toxicology Society, he has been a member of the Home Office Advisory Committee on Animal Experiments.

He has long had an interest in the history of science, and was Honorary Director of the Wellcome Institute for the History of Medicine.

# Man and Mouse

ANIMALS IN MEDICAL RESEARCH

*Second Edition*

**William Paton**

Oxford   New York

OXFORD UNIVERSITY PRESS

1993

Oxford University Press, Walton Street, Oxford OX2 6DP

Oxford New York Toronto
Delhi Bombay Calcutta Madras Karachi
Kuala Lumpur Singapore Hong Kong Tokyo
Nairobi Dar es Salaam Cape Town
Melbourne Auckland Madrid

and associated companies in
Berlin Ibadan

Oxford is a trade mark of Oxford University Press

British Library Cataloguing in Publication Data
Data available

Library of Congress Cataloging in Publication Data
Paton, William D. M.
Man and mouse : animals in medical research / William Paton.—2nd ed.
p. cm.
Includes bibliographical references and index.
1. Vivisection. 2. Medicine—Research. I. Title.
[DNLM: 1. Animals, Laboratory. 2. Ethics, Medical. 3. Research.
Q Y 50 P312m]
619—dc20 HV4915.P38 1993 92-16412
ISBN 0-19-286146-8 (pbk.)

1 3 5 7 9 10 8 6 4 2

Typeset by Best-set Typesetter Ltd., Hong Kong

Printed in Great Britain by
Biddles Ltd.
Guildford and King's Lynn

*To Phoebe*

*St Martin-within-Ludgate*
*22 August 1942*
*Perpetuae augurium harmoniae*

# Preface

In the nine years since the first edition was written, there have been several important developments that need to be taken into account. Most obvious, perhaps, is the growth in the activity of the Animal Liberation Front and its related bodies. It has prompted the preparation of an Appendix collating their history and actions for comparison with their claims.

There has also been new legislation, replacing the previous 1876 Act. The resulting pattern of activity is reviewed, and some account is given of the related changes taking place in Europe and the USA. This country remains unique for the high quality and detail of its supervision of animal work.

There are still some remarkable statements and recommendations made about biomedical experimenters and their work, which amount to gross misunderstandings of scientific work as a whole. It has therefore been felt important to retain the rather general chapters on the nature of experimental work and the pattern of discovery. 'Serendipity' is a word commonly stressed, although its origin seems less familiar. The text of Horace Walpole's letter has therefore been included, together with an account of the context of Pasteur's famous remark about 'chance favouring the prepared mind'.

There has been an increase in philosophical discussion, of a curiously circumscribed kind, dominated by a footnote in Jeremy Bentham's writing. It was felt desirable, therefore, to broaden the discussion of the ethical aspects: for instance, by including Rawls's theory of justice, Schweitzer's 'reverence for life' (with samples of his thought), and an Appendix surveying the biblical tradition (but not the later theology), which colours so much of the discussion. The reductionist nature of the alternatives campaign is also brought out. Since pain and suffering are at the centre of the debate, a fuller account has been given of our present physiological understanding of them.

New material on poliomyelitis, tuberculosis, and tetanus have been chosen, as illuminating the role that animal experiment actually plays, and the extent to which social factors, as against medical research, influence disease incidence.

Perhaps the most important developments have been in toxicity testing and in the use of *in vitro* methods for screening out or for initial classification of the more toxic materials. This work, with the important contribution made by the British Toxicological Society, has merited a new chapter devoted to toxicity testing. On the one hand, it has become clear that any procedure which is to allow a responsible regulatory body to declare some substance safe is bound to involve whole animal testing. On the other hand, a good many toxic effects are on systems whose genetic expression is common to a range of cell types in a range of species. Use of such systems *in vitro*, although representing only part of the whole genome, would serve to identify such toxins. They should also allow the design of any necessary subsequent animal tests to be done more precisely with less or no suffering. There have also been major advances in the technology for measuring the various parameters of an animal's physiology. All this suggests that a new goal—namely, that animal experimentation could come to involve no more suffering than is experienced by pets or other domestic animals—can now be seen as possible.

The first edition of this book ended by drawing attention to the irony that it was beneficiaries of medical research, the healthiest generation ever known, who were themselves harassing medical research and its workers. Today, one can end more positively. The life of animals is terminated in many ways—some of them extremely brutal and most of them carrying little of what one could call a purpose. But with the new possibility that where there is suffering involved in animal experimentation, it can be reduced to something comparable with that of pets and other companion animals, a new conception appears: of improvement in knowledge and future benefit to man and animals being seen as giving a new purpose to animal life, within a relationship sometimes already familiar: human trustee and animal companion.

W. D. M. P.

*March 1992*

# Contents

# Acknowledgements

I am grateful to my one-time employers the Medical Research Council for permission to reproduce Figure 5 from their 75th Anniversary report; to the Research Defence Society for Table 14; and to the *New Scientist*, in whose pages I first read Bob Andes's poem. I am especially indebted to friends who provided information that I could not get for myself: Professor A. Angel, Professor R. Chapman, Dr Tilli Tansey, Dr Iain Purchase, Dr Phil Botham, and Dr Jack Botting. Without their generous help and comment, the revision would not have been possible.

Perhaps, too, I could acknowledge, along with millions of others, a debt to some mice for my survival: those used by Domagk and by Florey in their chemotherapy experiments, some fifty to sixty years ago.

# List of Figures

# List of Tables

# Ghosts

Bob Andes

Aren't you
They asked
Haunted
By the ghosts
Of innocent mice
And rats
And cats
And, yes,
My God,
Those sad-eyed
Trusting
Dogs
That have suffered
And died,
That suffer
And die
Even now
In your damned
Laboratory experiments?
Aren't you
I replied
Haunted

By the ghosts
Of innocent men
And women
And boys
And, yes,
My God,
Those sad-eyed
Trusting
Little girls
Who have suffered
And died,
Who suffer
And die
Even now
Because I have not yet
Done enough
Damned
Laboratory experiments?

One should choose ghosts
Like friends.
Carefully.

# 1

# Introduction

The history of man's relation to the animal kingdom reaches back through the centuries, with the animal's role taking many forms, ranging from deity, totem, or scapegoat through companionship and co-worker to exploitation.[1] Debate about animal experiment is more recent, but is at least 150 years old.[2] One of the most interesting early documents on the subject is a paper written in 1831 by the neurologist Marshall Hall. Hall had been a pioneer in establishing the concept of the reflex arc in the spinal cord, and he invented a successful method of resuscitating the drowned. He also campaigned against slavery in the USA and against flogging in the British Army. In his paper he puts forward his own proposals for the regulation of physiological experiments, which were just beginning to throw a flood of light on our knowledge of the working of the animal body.

The first principle to be laid down for the prosecution of physiology is this: we should never have recourse to experiment, in cases in which observation can afford us the information required . . .

As a second principle . . . it must be assumed that no experiment should be performed without a distinct and definite object, and without the persuasion, after the maturest consideration, that that object will be attained by that experiment, in the form of a real and uncomplicated result . . .

It must be admitted, as a third principle . . . that we should not needlessly repeat experiments which have already been performed by physiologists of reputation. If a doubt respecting their accuracy, or the accuracy of the deductions drawn from them, arise, it then, indeed, becomes highly important that they should be corrected or confirmed by repetition. This principle implies the necessity of a due knowledge of what has been done by preceding physiologists . . .

. . . it must next be received as an axiom, or fourth principle, that a given experiment should be instituted with the least possible infliction of suffering . . .

Lastly, it should be received as a fifth principle, that every physiological experiment should be performed under such circumstances as will secure a due observation and attestation of its results, and so obviate, as much as possible, the necessity for its repetition . . .

In order fully to accomplish these objects, it would be desirable to form a society for physiological research. Each member should engage to assist the others. It should be competent to any member to propose a series of experiments, its modes, its objects. These should be first fully discussed—purged from all sources of complication, prejudice, or error—or rejected. If it be determined that such series of experiments be neither unnecessary nor useless . . . they should then be performed, repeated if necessary, and duly attested. Lastly, such experiments, with the deductions which may flow from them, may then be published with the inestimable advantage of authenticity.

Pursued in this manner, the science of physiology will be rescued from the charges of uncertainty and cruelty, and will be regarded by all men, at once as an important and essential branch of knowledge and scientific research.[3]

## An old debate in a new context

As one reads this, many of the issues discussed today can be recognized: consideration of alternatives, defined objectives, avoidance of repetition, minimum suffering, proper record, open criticism. These remain the objectives today. Are we then merely raking over the embers of an old fire, that sometimes dies down, sometimes flares up, over which men will never reach agreement? Or is there something new to be said? This book will argue that both these statements are true. The question of animal experiment depends in the end on the view man takes of himself and of the animal and inanimate world around him: it focuses on his choice between the avoidance of suffering now and the avoidance of suffering and remediable ignorance in the future. I do not believe there is any set of rules, any 'algorithm' suitable for a computer, which allows an unequivocal answer to the question, but rather that it forms part of the familiar pattern of moral decisions of everyday life about which men have differed, do differ, and will differ. But at the same time the debate today is in a different context from that in the past, and it is worth looking at some of the reasons for this.

One obvious fact, apparent to anyone who considers animal and human life in past centuries, or even decades, is a striking *rise in standards* of physical and mental well-being. Even as late as the 1930s, a schoolboy could have lost a companion from tuberculosis, mastoid infection, diphtheria, or scarlet fever, and might play with a friend crippled by polio. Deformity, pain, and disability were familiar experiences. For the routine pains of sinusitis, colic, peptic ulcer, or 'rheumatism', laudanum (or alcohol) was available for those who could afford it in the last century, and aspirin in this; but a great deal of suffering was simply accepted as part of life. Cruelty for its own sake seems always to have been re-probated; but parents who had lost most of their family in infancy, and learnt to speak coolly of it, could readily be equally cool about the suffering of animals. Today, however, in comparison, most people in the industrialized countries are healthy, straight-limbed, and pain-free; childhood ailments no longer threaten the bloom of childhood and adolescence—so rapidly lost in the past—and we can expect with some confidence a life-span of threescore years and ten in which, if disease intervenes, effective palliation at least will be available. The care of animals has improved in parallel, spilling over quite naturally, one could suppose, from man to his domestic companions and thence to animals generally. The same therapy, the same local or general anaesthetics, the same surgery, are used to abate their suffering as are used in man. Just as with man, for whom 'mental distress' may now be the cause of sub-stantial compensation in the courts, so the animal-lover broadens his concern from its focus on pain to include 'distress' or 'suf-fering' in a more general sense. It is important to recognize that the debate takes place in a context where standards have risen greatly, and must be expected to go on rising.

A second factor, bringing animal experiment more to the fore, is its *growth in scale*, now about 3.5 million procedures per annum, mostly on mice and rats. This needs to be put in context. It amounts to about 1 per head of population every 15 years or 5 mice in a lifetime. At the same time, hundreds of millions of animals, especially chickens, are killed annually for food, and several thousand tons of venison are sold each year. The Vegetarian Society has calculated that one Briton, in an average life, consumes 8 cows, 36 pigs, 36 sheep, and 550 poultry. Use of animals for hunting or to obtain fur continues. A remarkable report is that there are 10,000 people licensed to keep falcons and other birds of prey. Wild life, unseen by the town-dweller, continues with its own pattern of predators and victims.

Noteworthy is the growth of the population of dogs and cats, each of which has risen to about 7 million. Of these, around 200,000 are put down annually as unwanted or strays. A recent study of domestic cats exemplifies the scale of natural predation. It was found that on average each cat brought home annually an average of 14 items of prey, mostly birds, but also creatures such as woodmice, house sparrows, and voles. This gives a national total of around 100 million prey a year.[4]

Animal experimentation grew in past years, too, reaching a peak in 1976, and is now declining. Its growth was due to the discovery, essentially in the last century, of how effective animal experiment is (amongst other methods) in leading to great advances in knowledge and to equally great benefits for men and animals. That inevitably led to its *adoption by industry*. This was essential, for it is only industry that makes available to the many what would otherwise be available only to the few. As a result, the greater part of animal experiment, as estimated by the number of animals used, is now carried out by pharmaceutical industry.

At the same time the general rise in the standards of health, catalysed by consumer pressure, has reflected itself in the *growth of regulatory bodies* concerned to maintain and improve these standards. To do this, such bodies themselves lay down animal tests that must be done, not only on medicaments, but also on foods and chemicals used in the environment—indeed *any* chemical to which workers may be exposed, before it is allowed to be made generally available to industry. Such regulation is now becoming increasingly important at the European level.

In addition, the involvement of industry and of technological change sometimes enlivens the argument with more general issues: criticisms of capitalism, profit-making, interference with nature, modern medicine, even of scientific investigation itself.

One cannot hope, therefore, to achieve any finality of decision. Moral debate is inevitable, and one must also recognize that the material of the debate is changing. Standards will go on rising, both in the search for further improvement in human and animal well-being, and in the attempts to ensure safety in the way those improvements are brought about. Techniques will improve to meet these demands. Continued adjustment will be obligatory.

# The nature of the argument

There is another change to notice. A century ago, the two sides in the argument were fairly well balanced: there was not a large body of animal experiment; so the antivivisectionist could well familiarize himself with and genuinely understand a good deal of it, while the experimenter had time to answer the charges. Scientific journals such as *Nature* and the *Lancet* were more widely read. Literary journals such as the *Nineteenth Century* published scientific material. Laymen could, if not take part, at least align themselves with either side; and one finds bishops, lawyers, MPs, the aristocracy, and literary men lending their names freely one way or the other. So even if, from the start, the debate was polarized at the extremes, and charges of cruelty, ineffectiveness, and arrogance were liberally exchanged with those of misrepresentation, obscurantism, and a deeper cruelty, yet both the debate and its materials were reasonably open.[5]

But the situation changed. Scientific knowledge expanded enormously; it became more demanding of the investigator, and harder for the layman to understand. Biological and medical science, like other branches of science, became segregated as full-time professional careers. Science and medicine became institutionalized, so that an 'establishment' began to grow. At the same time, as the demands on the scientist grew, explanation by scientists themselves for laymen of what they were doing diminished: this was partly as a result of the growth of learned societies, which took the place of the British Association as the scientists' professional forum. Descriptions for the layman of scientific work by the scientist began to go by the board, apart from the activities of interpreters, 'popularizers of modern science' (the very phrase expresses the distance of actual scientific practice from the public's general understanding). Indeed, a good many medical scientists seemed to begin to feel that it was better to let the achievements of science and medicine speak for themselves, and for them, in modern jargon, to 'adopt a low profile'. So common ground for the materials for debate has been obscured. On the one hand, there are antivivisectionists and more or less radical animal welfare groups: most of their arguments are non-technical and easily accessible to any thinking person; they are well endowed by benefactions from sympathizers; and most Members of Parliament receive a regular mail from them. On the other hand, we see the large body of scientific and medical practice—together with the universities, Royal Colleges, and other institutions—and a range of major industries, both pharmaceutical and of other types, involved with animal experiment. These command considerable resources, yet make relatively little public contribution to the general debate: this has largely been left to one or two small groups speaking for the experimenter. The public is thus rather poorly informed about the great body of current and past scientific work, apart from accounts in the press of 'breakthroughs' from time to time.

The situation has been further exacerbated by the scientific inadequacy and prejudice of some animal welfare comment. One may read, for instance, that a leading animal campaigner believes that 'anyone with an ounce of medical knowledge would have known that a volatile anaesthetic doesn't get into the bloodstream',[6] although the fact is that it is only through the bloodstream that it passes from lung to brain; or from another such source, that it was 'almost without doubt' that the only experiments in the medical sphere were those involved with cancer research, diagnostic procedures, and drug research and that 'high-grade medical research work—the sort of work being done not in commercial laboratories but in universities—probably accounts for only a few thousand animals each year'.[7] Or one may find in a distinguished daily paper the same picture of a monkey, apparently smoking a cigarette, inserted repeatedly into articles on animal experiment in Britain, even *after* an enquiry was reported in the House of Lords which had shown that it is a Russian picture, with no explanation or reference, distributed by the Tass News Agency.[8] A further example is the bizarre misunderstanding of the LD50 toxicity test revealed in a much publicized report presented by the Committee for the Reform of Animal Experimentation (CRAE) to a government committee.[9] The scientist tends to react to such things with contempt, which is met with imputations of arrogance and comparisons of animal experiment with racism or Nazism, matching the comparisons with slavery made in the last century.

Another unhappy development has been the rise of physical violence by a wing of the animal welfare movement—a historical account of this will be found in Appendix 1. This inevitably reduces the willingness of individual scientists, already deeply engaged in teaching and research, to come forward to explain their position: too often doing so has led to personal or family harassment or other damage, or to harassment or physical damage to the institutions they work for.

A last event has been the entry of moral philosophers into the debate, partly as a result of the professional philosophical interest in the questions presented by man's experience of pain.

The result of all these forces is a fairly voluminous animal welfare and antivivisectionist literature, including some substantial books on animal rights and animal welfare. But there is a distinctly thin literature on the animal experimenter's position, to explain both the benefits achieved and how it was done. There is Sir Leonard Rogers's small, trenchant, but not generally known volume *The Truth about Vivisection*, published in 1937, and Lepage's *Conquest* published in 1960. Smyth's *Alternatives to Animal Experiments*, commissioned by the Research Defence Society, appeared in 1978, and there are now some valuable papers (especially Uvarov (1985) which provides a useful general review from the point of view of a distinguished veterinary surgeon, including past and present legislation and history and recent advances, especially in veterinary vaccines). The lack of larger studies may be because medical scientists prefer, or feel they ought, to devote their efforts towards new discovery rather than history. Other factors are that for historians, any flavour of 'triumphalism' is now unfashionable (outside advertising), and social and personal history are rather more generally accessible to historians (and perhaps more palatable for publishers) than the technicalities of research.

The present book is an attempt to go some way to redress the balance. The pattern of it is in part dictated by the views mentioned above. Because the nature of biomedical work and its achievements have been poorly explained, some attempt will be made to do this. Dogmatic statements about 'great advances' and 'breakthroughs' have seemed to me, in public discussions, almost totally useless. But particular instances, with an adequate explanation of the specific scientific issues involved, of the research opportunity or requirement that prompted the work, of the procedures used (and the reasons for their choice), and of the outcome in new knowledge and benefit, seem consistently to arouse interest and respect, even in those who remain opposed.

It will also be necessary to spend some time on general arguments, although the approach generally will be that there are no 'knock-down' arguments, that moral philosophy may clarify but does not prescribe, and that in the end it is a matter for collective moral decision. The layout is not strictly logical, but is that which has seemed to arise naturally in argument, to a large extent centring round particular recurrent questions. A number of special issues will be taken up in separate sections.

Finally, although there can be deep feelings about the issues, it seems clear from the past that too vigorous expression of personal attitudes merely deepens existing entrenched positions. Any writer is bound to have his own views, and these cannot be disguised. The view taken here, however, is that an argument gains no special validity because a particular individual holds it (or opposes it) or because of the strength of his opinion. The objective is to lay out as much of the argument and evidence as possible in a generally accessible manner, so that readers can judge for themselves.

# 2

# What is included under the word 'animal'?

> For there is in this Universe a Staire, or manifest Scales of
> creatures, rising not disorderly, or in confusion, but with a
> comely method and proportion: between creatures of meer
> existence, and things of life there is a large disproportion of
> nature; between plants and animals or creatures of sense, a
> wider difference; between them and man a far greater; and
> if the proportion hold on between men and Angels, there
> should be yet a greater.
>
> Sir Thomas Browne, *Religio Medici* (1643), section 33

Our first choice of meaning might well be like that made by
George Orwell in *Animal Farm*: Boxer the horse, dogs, pigs,
sheep, cows, goats, donkeys, the cats, ducks, geese, pigeons,
turkeys, the ravens—even rats (voted in, by the others, with dog
and cat dissenting).[1] A restriction to non-human vertebrates, particularly the warm-blooded, certainly corresponds to one pattern
of use. But the term can be used much more widely. The discovery of the microscope 300 years ago opened up a new world,
and Leeuwenhoek used the term 'little animals' for creatures
as small as protozoa and microbes.[2] Albert Schweitzer, with his
reverence for life, records that he wondered, as he looked down
his microscope, if he had been justified in killing the micro-
organisms displayed there.[3] What are we to say of the frog, the
fish, the bee, the ant, the wasp, or the lobster? Or does 'animal'
simply mean anything not a vegetable or a mineral, and is that the
range of creatures that we wish to provide for?

Once we broaden our view in this way, it is obvious that our
concern is not with the semantics of the word 'animal', but with a
deeper issue. As we survey the whole world around us—and we
can consider nothing less than that, if we think of the range
of scientific enquiry and practical use—can we see lines of

demarcation? In past centuries such a panorama was implied in the idea of a Chain of Being, *scala naturae*, a Ladder of Perfection, stretching from the mineral world through lower forms of life up to man, and then to purely spiritual existences: 'There is in this Universe a Staire', wrote Sir Thomas Browne, 'rising not disorderly, or in confusion, but with a comely method and proportion.'[4] We can envisage such a panorama today, although that 'Staire' takes a different aspect. It is no longer seen as a smooth development from one creature to another. Today we recognize that each species itself stands as the latest expression of one of a vast number of evolutionary branches. Evolutionary relationships might, indeed, help us in drawing lines of demarcation; but we also know that the same function—say flying, swimming, or seeing—can be achieved by creatures following totally different evolutionary pathways. So we cannot in general rely on finding some evolutionary branch-point that will decisively mark the point of appearance of those characteristics in which we are interested.

## The biological continuum and operational decision

Indeed, what is it in which we are interested? One demarcation at issue is an upper bound: namely, whether man is to be distinguished from animals and the rest of creation. But a lower bound is needed too. We do not propose to legislate for cruelty to clay or gravel or, one imagines, for amoebas or cabbages. So where do we place the line where our particular care about experimental work stops? In question, too, are the criteria by which we identify such demarcations. There are many proposals; and both they and the difficulties they bring are conveniently illustrated by looking at some of the links in the chain of being. Even in the inorganic world we note that some things, like crystals, grow. Moving on to the organic world, we find that viruses, sometimes consisting of only a few different (though large) molecules, can reproduce in a suitable host cell. Bacteria can reproduce, both sexually and asexually, swim, and emit noxious substances. In the vegetable kingdom, flowers move, turning to the sun. The clematis curls its prehensile tentacles round the twigs or wires it climbs up. The sensitive plant *Mimosa pudica* responds to sound, folding its leaves at the noise of a handclap. Insectivorous plants such as the sundew (*Drosera*), as Charles Darwin showed,[5] are exquisitely sensitive to chemicals—one thousand millionth of a gram of ammonium acetate being enough to make one of its hairs bend. Moving on to the invertebrates, we notice that even so simple a creature as the jellyfish has a nervous system and can respond to stimuli. If we watch an ant or a bee or read about the powers of an octopus, it is hard not to admit the words 'learning' or 'purpose' in some sense in describing their behaviour. The leech and the snail[6] contain receptors for pain-killing drugs, and fish are rich in opiate-like substances (peptides) similar to those in our own brains. There is no need to elaborate on the range of behaviour possible for cold-blooded or warm-blooded vertebrates, ranging from the frog through birds and primates to man. I do not think anyone has found a sharp, rigorous point of division in searching for our lower bound, between 'animal' and 'non-animal', whether one considers growth, movement, reproduction, self-protection, response to stimuli, sensitivity, learning, apparent purposiveness, or relationship to pain mechanisms familiar in man.

Likewise for our upper bound. We do not need to *prove* that animals other than man can suffer or solve problems or experience consciousness or fear. It is enough to note that no reason exists why they should not share these capacities, and that as our knowledge of animal and primate life develops, the more frequently vestiges of behaviour are recognized that can plausibly be compared to the human experience of pain, apprehension, sorrow, frustration, guilt, altruism, laughter, symbolic communication, thought, or invention. The behaviourists can, of course, formulate models for such behaviour—after all, machines (and computers, which are a type of machine) can be designed to learn and solve problems; and it will probably never be possible to exclude the possibility that behaviour, however complicated, is ultimately describable as the sort of machine that the French philosopher Descartes conceived an animal to be 300 years ago. But this seems equally unlikely ever to be provable. So we need to think of the panorama of creation with its millions of species—each species itself blurred by varieties and normal variability within the species—as presenting a continuous spectrum, with man and his characteristic qualities included.

To find such continuity and a lack of dividing lines is a very common human experience. For instance, there is a continuous gradation between, say, lightness and darkness, with no divisions.

Yet in practice we can distinguish the two ideas and use the words 'light' and 'dark' effectively, as in specifying the darkness of a photographic dark room or the lightness of a workplace or street at night. More painfully, perhaps, we set levels in the continuum of income at which taxation changes, a level of blood alcohol that is liable to prosecution, levels of scholastic achievement to gain particular qualifications, and levels of speed, distance, or height that athletes must meet to qualify for competition. A great deal of human life consists in *operational decision*: that is, solving the problem of drawing a line at some point on a continuous scale, in order to allow some type of decision to be practicable.

How this is done would be a major study in itself. But a simple example, that of the division of the 24 hours into 'night' and 'day', points to two obvious processes. The first is to decide that the divisions should be made around sunset and sunrise, because these are the times at which the rate of change from one phase to the other is fastest; this narrows the zone of potential ambiguity. Then the final, detailed definition, where it is necessary, can be made to depend ultimately on a general agreement on some acceptable and practicable defining procedure—for instance, using astronomical data, establishing time zones, or taking account of industrial needs by introducing 'daylight saving'.

## The upper bound: man–animal

So we return to consider our upper and lower bounds for the 'animal' world, admitting continuity, but asking now whether there are practical, operational, generally accepted grounds of demarcation. As to the upper bound, I believe there is in fact a general consensus recognizing a major operational difference between man and other vertebrates: *the main difficulty is in articulating the nature of the difference*. The parallels in bodily physiology, in the nervous pathways transmitting signals from damaged tissues, and in the behaviour in response to such signals make it improbable that man is grossly more sensitive to 'pain' than other warm-blooded vertebrates. In fact, in acuity of some sensory perceptions he may be inferior: the dog's sense of smell, for instance, is far more acute than man's. But there is one general area where man is outstanding, not often pointed out in this context: that is, in *his capacity to accumulate and share his experience*

*and his solutions to the problems he has encountered, by means of the
spoken and particularly the written and printed word.* This means that
we do not restrict ourselves to assessing 'capacities', but look also
at actual *achievement*.

One can recognize that other beings can, in a sense, accumulate
their past: a coral reef or an anthill are examples, but these are
mere aggregations of the products of the same activity. The process
of evolution might also be construed as a process of accumulating
solutions to environmental pressures. The process is slow, however, spreading over the millennia. One could argue, too, that the
rough generalization that 'ontogeny repeats phylogeny' shows the
accumulated record of our evolutionary past—for instance, in
the vestiges in our bodies corresponding to gills or a tail. Yet these
vestiges are functionless, and do not give us the capacity to breathe
under water or to swing with limbs free from a tree, in the way
that man's past discoveries can continue to be put to his service.
Much evolutionary change in fact represents the exchange of a
less advantageous capacity for a better one, and much of the past
history is deleted.

But at best, all that we can discern are the merest vestiges of
the beginnings of those qualities which led to man's accumulative
capacity; and they are trivial compared to the consequences of
that capacity. At the heart of it is not merely accumulation but
communication. This is particularly striking in medicine. The
possibility of human anaesthesia by ether was discovered in 1846
in Boston, and within a few weeks it was being practised across the
Atlantic. At once, new anaesthetics, improvements in technique,
new applications began to flow; and the fact that surgery could
now be painless opened the way in its turn to antiseptic and
aseptic techniques. Röntgen's X-rays spread with a similar rapidity.
The manner in which each of us builds on the results of other
people's work is peculiarly vivid to the scientist, and is crystallized
in the bibliography appended to his papers, the eagerness with
which he reads his journals, the proliferation of abstracting
journals (and even abstracts of abstracts). More generally, the
proliferation of textbooks, instructional handbooks—indeed, the
whole educational system—expresses the same establishment of
past achievement as a stepping-stone for new achievement to
come.

The effects of accumulated knowledge are so familiar, so much taken for granted, that they are difficult to articulate. They are evident in the diversity and extent of man's artefacts, his technological power, but above all in the records of his activity—in the libraries of the world (and soon in the computer stores). It is because of that written record that the schoolboy can master in a few weeks the calculus that Archimedes began with his 'method of exhaustion' and whose formulation taxed the greatest brains of the seventeenth century. The musician can see the development of Pythagorean harmonies, read Rameau's treatise, explore the works of Bach, and imagine a tutorial with Hindemith, before he strikes out on his own. The naturalist does not start, like Adam, naming species from the beginning, but draws on the descriptive work of thousands, as well as on the classificatory guidance of Linnaeus and his successors. The engineer need not invent steel or concrete, or the steam, diesel, or jet engines afresh, but can envisage yet new materials or new ways of harnessing energy. The outcome, of course, is that man has an incomparably greater knowledge of the world than any other species, and to match this an incomparably greater power, and prospects of yet more. Whatever vestiges there may be in animals of the qualities that have allowed all this, the fact is that they are fulfilled only in man.

It is worth stressing that the argument is not so much about capacities—whether animals do or do not share some particular ability—but about *achievement*, by which the significance of a capacity can be judged by its actual outcome.

It is, in fact, curious that man's capacities are so rarely considered in the way that those of animals are. The thinking often starts by considering some animal capacity, and then seeing whether man can match it—which, not surprisingly, with a different ecological niche, he often cannot. Adam Smith, the economist, is not usually invoked in this context, but he exemplifies the method of starting with man, for once. He is famous, if not notorious, for his doctrine of the division of labour, free competition, and the 'silent hand' that transforms general self-interest into general benefit. Less quoted is the passage in the second chapter of his book *An Enquiry into the Nature and Causes of the Wealth of Nations*, where he explains how it all starts with the human propensity to a particular type of *co-operation*—namely, the propensity to 'truck, barter and exchange one thing for another'. 'It is common to all men, and to be found in no other race of animals. . . . Nobody ever saw a dog make a fair and deliberate exchange of one bone for another with another dog. Nobody ever saw one animal by its gestures and natural cries signify to another this is mine, that yours: I am willing to give this for that.'[7] It appears to be accepted by the animal 'rights' philosopher T. Regan also that human beings alone of 'terrestrial creatures' can enter into a contract.

As well as language and writing, many words have been put forward as expressing a specifically human character, some more seriously than others. They include 'art', 'blame', 'concept of the future', 'consent', 'cooking', 'making promises', 'moral sense', 'numbering', 'politics', 'punctuation', 'reason', 'sense of beauty', 'sense of another mind', 'religion', and 'tool-making'. But if one looks at the extraordinary variety of human achievement, one must put high the 'ability to co-operate by means of fair exchange' in sharing out the work to be done. It reveals itself in every part of human life, whether 'developed' or 'developing'.

One must remember, too, that man constitutes the first species that has begun to understand its own evolution. He can therefore begin to find ways to free himself from past determinisms, and begin to frame his own future path.

## The lower bound: vertebrate–invertebrate

Where do we place the lower bound? This has always presented difficulty. From the start, it was the animals that man had domesticated or made his companions, especially the dog, cat, and horse, which aroused especial concern; and from this arises general agreement that all warm-blooded mammalian vertebrates must be included. Such was the feeling in 1875–6 about the chosen favourites that the initial proposals were to ban absolutely any experiment on cats, dogs, or equidae (horses, donkeys, mules);[9] but in the final legislation, experiment on these was allowed under special certificates. At that time experiments on rats and mice were fairly limited. But soon rats and mice became the commonest subjects for experiment, and it was recognized that the arguments that they can suffer are no less strong than they are for cats and dogs: this has led to more attention being paid to them. Argument continues, however, about how far domestication or biological evidence of the extent of brain development should influence the level of concern. Yet more difficult has been the category of cold-blooded vertebrate animals, particularly frogs and fish. It might indeed be the case that no form of consciousness is possible at these lower temperatures: brain development certainly falls far short of that in the mammal. Yet there remain sufficient similarities, for instance in the details of brain structure, for the contrary to be argued. The original draft of the 1876 Act excluded 'cold-blooded' animals from its provision, and it was only in the committee stage in the House of Commons that a new distinction—that between 'invertebrate' and 'vertebrate'—was adopted.[10] This meant that not only the frog, but also the tadpole, newt, and even the smallest fish, are all included under the Act: it was therefore a reasonably cautious dividing line. It was also one that the Littlewood Committee reconsidered in 1965, and saw no reason to change.[11] Some would think inclusion of tadpole or newt unnecessary, others that the range should be widened to include some invertebrates such as the octopus or lobster.

The centre of the problem appears to be the conditions required for consciousness, particularly consciousness of pain. Of course, one might envisage animals existing into whose consciousness pain never enters. But this seems improbable when one reviews the physiology of pain, particularly the way it warns of body damage and commands attention by overriding other sensations. It is discussed more fully in Chapter 4. Here it will be assumed that any animal with consciousness would in fact be aware of 'nociceptive' (i.e. harm-producing) stimuli. Their significance is not that they produce reactions, but that they are *felt*. The essential question reduces, therefore, to the conditions required for the presence of consciousness as such. It was the ability of anaesthetics to remove *consciousness* of pain that made modern surgery possible.

Certain main ideas seem to underlie people's thoughts as guides to the probability of some state of consciousness and thus potentiality for suffering: responsiveness, level of complexity of behaviour, analogies to man in nervous-system structure, and size of nervous system. Some examples may serve to show how inconclusive such ideas can be. First, the highly complex behaviour of the ant, bee, or spider is associated with a tiny nervous system, which few would believe capable of consciousness; it is clear that complex behaviour, whose neural substrate we may well not yet fully understand, is not sufficient evidence. Secondly, consider a man who has suffered a transection of his spinal cord. He has now lost all consciousness of the parts of the body supplied by the cord below the transection. But after 'spinal shock' has passed off, that severed piece of spinal cord is still capable of mediating a considerable range of responses to stimuli. Under the microscope it would be found to consist of a considerable mass of nervous tissue, displaying very great complexity of connections between the nerve cells. Despite these reactions and this complexity, no one, I think, would suggest that the transection has created another 'consciousness' centred in this separated piece of cord. Thirdly, consider an anaesthetized man. Consciousness is lost, and suffering cannot occur; yet a wide range of reflexes still operate; response by movement still occurs to painful stimulation; brain waves can be recorded; and with light anaesthesia seemingly purposive movement can be displayed. Analogous phenomena are seen, of course, in a bird or frog deprived of its brain. Such examples show that neither responsiveness to stimuli nor complex behaviour nor simple mass or complexity of neuronal tissue are enough to determine whether consciousness is present or not.

One particular case has drawn attention, that of the cephalopod—for example, the octopus or squid—especially because of the learning responses it can show with visual stimuli. But now a simple computer program, with only eight mutually connected elements ('neurones'), has been shown able to 'learn' to recognize a simple shape.[12] One needs, therefore, to be more critical. One should note too that there are more neurones (350 million) in the arms of an octopus, each able to crawl if cut off, than in the whole of its central nervous system (134 million).[13] How many consciousnesses should one reckon?

The multiplication of neurones will bring another problem, that the time needed for synaptic action can limit the scope of operation. Speed of operation already limits computer capacity. Now, synaptic physiology becomes considerably faster as body temperature rises. It is not impossible that it was the rise in body temperature among vertebrate species that allowed the higher nervous functions to evolve. In man, anaesthesia follows if the body temperature falls more than 10°C to below 30°C. A similar fall probably provides the best way of anaesthetizing an octopus.[14] Perhaps the distinction between cold- and warm-blooded animals, suggested in the debates at the end of the last century, is a sound one.

If we try to look for some unequivocal basis for consciousness which we could use as a test for its presence in an animal, for the time being at least, we look in vain. And while it is a tame conclusion, we can conclude little more than that the answer lies in some particular pattern of organization. It may indeed be that for that pattern to be possible, a minimum number of nerve cells, or sets of nerve cells arranged in co-operating groups is required, just as a computer cannot perform particular functions without memory, programs, and data banks of a sufficient size. Thus there could well be a minimum brain size for consciousness to appear. But, as with a computer, that size will be necessary but not sufficient, and the function will be possible only when its organization is appropriate.

As we review such data, it seems that we have to proceed in a tentative way, aggregating our knowledge of brain structure and function, behaviour, and relationship to other animals as best we can. In trying to decide what to include in the term 'animal', therefore, we must conclude that, for our present purpose, the upper bound falls short of man, and the lower bound is placed at the division between vertebrates and invertebrates. It must be restated that this does not rest on any rigorous logical division, but on operational 'cuts' in an apparently continuous spectrum such as we are constantly making in other walks of life. The distinction is important: those who wish to argue that the upper bound is irrational, because no one can point to an absolute dividing line between man and the other primates, must go on to argue that the lower bound is equally irrational and equally has to be abolished. On that approach, if animals were to be treated as though they were man, then insects, and perhaps even plants, bacteria, or clay, would have to be treated as though they were animal.

## Discrimination between animals

The defining of the meaning of animal leaves the question of differentiation between them, mentioned briefly above. One source of such discrimination occurs, sometimes with attractive results, when a particularly careful study is made of some individual species. This invariably brings to light hitherto unsuspected capabilities, seeming to raise that species above its neighbours. Yet we should recognize what could be called a *'fallacy of exaggerated attention'*; it would not be accurate to set that species above its neighbours unless the same devoted attention were exercised throughout. The difficulties of discrimination are obvious, as between, say, a rat and a rabbit. Yet few would have difficulty between a chimpanzee and a tadpole. Many feel, too, that the relationship set up with domestic, 'companion' animals should be recognized; others, that even if we are fonder of cats and dogs, the rat and mouse can suffer equally. Bentham's question 'Can they suffer?' may still be asked. There is no simple answer, but a general approach could be to think in a general way of 'lower' and 'higher' forms of animal, just as (to use the previous analogy) we think of 'lighter' and 'darker'.

Because of the precision required in legal drafting, it is hardly possible to include statutory provision for the rough-and-ready idea that experiments should, where possible, be made on 'lower' rather than 'higher' forms. But Article 7 of the European Convention, which it is hoped will produce a general raising of standards internationally, offers scope for this approach:

When a procedure has to be performed, the choice of species shall be carefully considered, and where required, be explained to the responsible authority; in a choice between procedures, those should be selected which use the minimum number of animals, cause the least pain, suffering, distress, or lasting harm and which are most likely to provide satisfactory results.[15]

This is a wisely framed prescription. It carries us away from semantic problems back to the essential objective: to try to choose that animal with which the suffering will be least in the obtaining of satisfactory results.

There is one last point which may provide some reassurance to those uneasy about incorrect choices. In Britain, the welfare of animals is also covered by a variety of other legislation, particularly the Protection of Animals Act of 1911, which brought together a considerable body of previous legislation.[16] Its range is considerably wider so far as species go, but it differs from the 1876 Act in that it depends on *actual suffering having occurred*. (See Appendix 2, Clauses 1 and 15.) It is not always understood that the 1876 Act and its 1986 successor are fundamentally preventive, prescribing many conditions so as to avoid or minimize suffering, and that conviction for an offence could occur without any actual animal suffering taking place.

One hopes that this aspect of legislation regarding animal experiment will always be maintained. But the existence of the more general 1911 Act is valuable as a further protection, particularly relevant to any inadequacy in the placing of the lower bound in legislation dealing with animal experiment. The 1911 Act is there to deal with unjustified suffering in animals generally.

## Summary

1. If we review the 'scale of creation', it presents itself as a continuum in which we can discern no *absolute* dividing line separating creation into different categories, whether we use movement, growth, reproduction, self-protection, response to stimuli, sensitivity, learning, apparent purposiveness or evidence of pain-sensing mechanism, or evidence of suffering, problem-solving, consciousness, or fear.

2. Yet we readily acknowledge that our concern for the well-being of a cat or a dog is different from that for an ant, grass, or clay, and that the problem is a familiar human task of making operational 'cuts' in a continuous spectrum.

3. An 'upper bound' can be made between man and other animals, because of man's capacity to accumulate experience, to build on it, and to share it by means of the spoken and particularly the written and printed word. It is this that has given him a greater knowledge of, and power in, the world than any other species, and prospects of yet more. The vestiges of the qualities that allow this may exist in animals, but the fulfilment is only in man.

4. A 'lower bound' is commonly assigned either at the distinction between warm-blooded and cold-blooded creatures or between vertebrates and invertebrates. The former is perhaps more probable. It is argued that any being that can move and interact with its environment will have some system warning it of bodily damage, and that this warning will be present in any consciousness it experiences; for it is impossible to believe that a creature which possessed a consciousness which was *not* responsive to adverse circumstances would survive in evolution. The essential

question, in trying to place a reasonable lower bound, may then be reduced to estimating the extent (if any) of such consciousness.

5. If this operational approach is *not* adopted, there appears to be no reason for distinguishing between the care due to minerals, bacteria, vegetables, animals, and man: all should be treated alike.

# 3

# What is an 'experiment'?

## Observation, intervention, and analysis

It is vital to realize that we are discussing a process of discovering something that, before the experiment, was not known. It is this that provides the great objection to any legal provision which requires the beneficial outcome of an experiment to be precisely specified. We need to be clear about the nature of experimental work. It is not the same as new 'discovery' in the widest sense of the word, for that goes beyond the scientific experiments we are here concerned with. For instance, although if you have solved an anagram or found the roots of some awkward equation, you have discovered something you did not previously know, yet it did not involve experimental work. Nor is the world of experiment simply synonymous with that of science; much scientific work is purely observational, and it is sometimes urged that the animal experimenter (or medical research worker) would do better to *observe* nature more and *intervene* less. To illustrate these points it is worth considering a simple example of scientific investigation.

(1) We wish to know how two new sedatives compare in the duration of the sleep they produce. Two groups of six male mice are selected, weighing 20–1 grams, and each is injected with a suitable dose intraperitoneally (into the abdominal cavity). They are placed in a warm environment, and the time is measured at which the ability returns to right themselves within 10 seconds if turned on their backs. The difference in the average sleeping times of the two groups gives the desired comparison, and a statistical test (using the standard deviation of each average) allows us to assess the reliability of the test.

(2) Following further study of their satisfactoriness, these sedatives are to be compared in hospital patients having dif-

ficulty in sleeping. Patients weighing 50–1 kilograms cannot be ordered from a supplier, so we depend upon the chance of suitable patients being admitted to hospital, seeking to obtain two groups, made up of similar age, weight, and sex distribution, to each of whom a suitable dose is given. Instead of testing their righting reflex, nurses record their sleeping times, and the patients report their own estimate (often very different).

(3) It is known, however, that mice normally sleep by day, and their feeding and social activity takes place at night; so the first experiment can be regarded as incomplete, and a picture of the diurnal activity of these particular mice is needed. A number are set aside, therefore, of the same type as before, in a cage with adequate food and water; and an 'activity meter' is arranged (e.g. a radar field which records every movement, adjusted so as to reject respiratory movement but to record anything more extensive), and left running day and night for a few days, with a continuous recorder. This gives a graph of the pattern of activity for this group. Suitable times of day can now be chosen to repeat the earlier tests.

(4) Stimulated by this, one might wonder about the pattern of sleep in hospital patients. They are unlikely to be willing to be placed in a group of six in an enclosed space, with food and pellets, for a few days. So the nurses could simply be asked to record their sleeping behaviour without any comment or intervention. A series of very varied patterns results, which may be described by an average or by ranges or simply presented as a collection.

These four examples help us to escape from semantics, and to tease out some essential features. In each case there was a 'question' to be answered. All the procedures were 'observational', so that this word alone is not very instructive. They all produced something that was not known before.

But the four runs differ in three important respects. First, in (1) and (3) we deliberately *intervened* by setting up an experimental procedure and its raw material, *selecting subjects* (mice) by *particular criteria*; in (2) and (4) nature was allowed (by the chance of admissions to hospital of a series of individuals) to provide the raw material, and our role then was not that of intervention but only of selection. Secondly, in (1) and (2) we intervened by giving a treatment, whereas in (3) and (4) we only observed. Thirdly, in (1) and (3) a *comparison* was made between two types of treatment; in (2) and (4) we ended simply with a pattern, a description.

These runs, elementary as they are, fail to do justice to the richness and diversity of scientific work, but they roughly straddle the extremes of scientific study, from simple observation of 'nature's experiments' at one extreme to various forms of inter-vention at the other: providing the test objects, formulating an experimental procedure, and introducing some treatment or other 'perturbation' (to use a physical term).

It was the introduction of deliberate intervention into the study of natural phenomena, as opposed to observation, which so inspired the investigators of the seventeenth century onwards, and made them talk of 'the way of experiment'. Galileo, Torricelli, Gilbert, Boyle, Hooke, and their peers showed what a wealth of new knowledge could be gained if, instead of just observing natural events, you set up your own test to make events happen under your own control, choosing materials, procedures, and comparisons. That a springy material bent further if more heavily loaded must have been noticed by most people over the centuries; but it needed Hooke to compare the magnitude of the movement with the magnitude of the load (*Ut tensio sic vis*) to reveal the simple law. Similarly, once methods of producing a relative vacuum had been worked out, Boyle might simply have enumerated all the effects on objects placed within it; and indeed, to some extent, that is what he did. But he also compared, for instance, how volume increased as pressure fell (as others did conversely by testing how much a volume decreased when subjected to greater depths of water), revealing a very simple relationship, Boyle's law, that pressure multiplied by volume is a constant. The difference between random observation and deliberate questioning is illus-trated by his placing of a bell, struck by a mechanical device, within the evacuated chamber of his 'engine'. Why should he do this? It was to test whether air was responsible for carrying sound. The gradual fading of the sound and its return when the vacuum is released is one of the most simple and satisfying of his experiments.

# Holism

This concept of an intervening, deliberate simplification is of importance today, when biological and medical science are accused of being too analytic, of not thinking 'holistically', of ignoring the rich complexity of particular individual instances in natural life. If one turns back to Hooke and Boyle, can anyone believe that either could have arrived at his laws by mere inspection of the casual bendings or movements of the world around him? Nor does the point stop there; in fact, both laws are inaccurate and incomplete. The tension of a spring also depends on temperature and on the *rate* at which it is stretched; and the change in volume of a gas, caused by change in pressure, in Boyle's experiments would also have been sensitive to temperature, to the particular gas used (at the extremes of pressure), to gas dissolved in the walls of his containers or lost by diffusion, to water vapour forming or condensing, and many such factors. But until the first primary law was discovered, not even a start could be made in disentangling these other factors (leading in turn to the wider Van der Waals's law, and to new knowledge of intermolecular forces, to Henry's law of gas solution, to Fick's law of diffusion, and to the discovery of water-vapour pressure, dew point, and hygrometry). After centuries of observation, the discovery of how much could be done by simplification in analytic experiment constituted an intellectual thrill whose impact can still be felt in the writings of the time. Biological and medical events are far more complex than these relatively simple phenomena; and the more complex they are, the less hope there is that mere inspection can reveal any pattern and the greater the need for analysis. The complexity of such phenomena merely reinforces the need for experimental analysis, for disentangling the various processes and their interaction. The over-stressing of the individuality of natural phenomena leads to a blind alley. If there is *no* pattern, *no* generalization, *no* common principle to be found, then each case or event becomes isolated, bearing no relationship to any other. Individuality is, indeed, then at a maximum; but the price is that no individual can profit by the experience of any other—there are only special cases.

## The unknown and the unexpected

There is an extra dimension to the possible outcomes of an experiment. What happens quite commonly is not an answer to the question put, but something quite unexpected. An example would be to find in the experiment outlined above that one of the new sedatives did not produce sleep but excitement or convulsions (as is the case for one or two of them). A remarkably similar instance is that of a simple chemical derivative of ether, the classical surgical anaesthetic—namely, hexafluoroethyl ether—which proved to be so reliable a convulsant that it was used clinically as an alternative to electroconvulsion therapy. We should not be surprised how often this occurs, for it would only be if we had a full prior understanding of all the mechanisms in play in a living organism that we could be sure that we had put the right questions. It is, in fact, the mark of a shrewd investigator that he is on the look-out for the unexpected and gives it a chance to appear and be recognized.

Another example is worth recalling. As a young man, H. H. Dale (later Sir Henry) went into the pharmaceutical industry, and was asked by his employer, Henry Wellcome, to look into the pharmacology of a traditional remedy for inadequate uterine contraction in or after childbirth—namely, ergot, a fungal growth on rye.[1] Part of this work involved checking samples for their activity. Such natural material is liable to bacterial contamination, and as a result may contain a variety of products of bacterial metabolism (as does also, for instance, cheese). Although these contaminants were not related to the active principle of ergot itself, which was being tested on the blood pressure of anaesthetized animals, their effects—varying from sample to sample—aroused Dale's curiosity. The dividend was remarkable: first, some substances resembling adrenaline were found, which gave rise to great advances in our understanding of the sympathetic nervous system; secondly, a substance called histamine was identified, which opened up other fields of work: namely, in circulatory shock and what came to be called allergy. Thirdly, a substance called acetylcholine was discovered. This attracted Dale's special attention because he found that, after giving a dose that had apparently produced a fatal fall in blood pressure in an anaesthetized animal, the blood pressure nevertheless recovered fully in a few minutes. He reasoned that to be disposed of so quickly there must be something in the body equipped to destroy it, so that acetylcholine might have a deeper significance, as indeed was the case: it proved to be the first neurotransmitter whose role and identity was unequivocally established, and is now known to mediate the nervous control of our muscles, secretory glands, and important pathways in the brain. Thus a rather *ad hoc* practical question about ergot opened up three new areas of physiology and pharmacology.

It would be wrong, however, to picture successful research as commonly resting on pure luck. Reference is often made to 'serendipity' in this context, though its origin is less cited. It comes from the eighteenth-century writer Horace Walpole, describing in 1754 a literary discovery he had made, in a gossipy letter to Sir Horace Mann, ambassador in Rome:

This discovery, indeed, is almost of that kind which I call *Serendipity*, a very expressive word, which, as I have nothing better to tell you, I shall endeavour to explain to you: you will understand it better by the derivation than by the definition. I once read a silly fairy tale called *The Three Princes of Serendip*: as their Highnesses travelled, they were always making discoveries, by accidents and sagacity, of things which they were not in quest of: for instance, one of them discovered that a mule blind of the right eye had travelled the same road recently, because the grass was only eaten on the left side, where it was worse than on the right—now do you understand *Serendipity*? One of the most remarkable instances of this *accidental sagacity*, (for you must observe that *no* discovery of a thing you *are* looking for, comes under this description,) was of my Lord Shaftsbury, who happening to dine at Lord Chancellor Clarendon's, found out the marriage of the Duke of York and Mrs. Hyde, by the respect with which her mother treated her at table.

Such accidental sagacity, added to the active curiosity that Walpole also describes, is certainly displayed in successful research. But to it must be added what is probably a more important quality, training. Pasteur recognized this in his remark that 'chance favours the prepared mind'.[2] It is significant that his speech was made at a university inauguration, stressing the importance of practical work in science. The mind can be trained to be ready for what chance may present to it.

## 'Fundamental' and 'applied' research

Within experimental work, the distinction between 'fundamental' (or pure or basic or academic) and 'applied' (or practical or useful or mission-oriented) research can become important. For it may carry the implication that fundamental research is regarded as merely 'wanton curiosity', while it is only 'applied' research with practical benefits in view that could readily justify animal experiment. The distinction has become more common since its articulation in the Rothschild Report of 1971.[3] There, 'fundamental' work was regarded as simply adding to knowledge, whereas 'applied' was characterized as suitable for a detailed contract between investigator and customer. It was suggested that government support for fundamental work might reasonably amount to 5–10 per cent of the funds available, the remainder being suitable for contractual handling. The 'customer–contractor principle' may owe some of its attractiveness to its apparent definitiveness. It seems easier to judge whether a contract would ever be agreed for the undertaking of some investigation than to decide whether it is 'applied': being a contract, the 'use' of the work would be defined in some sense. It was therefore a proposal that appealed to administrators and to those seeking to make science meet their standards of 'accountability'. But it also ran across the grain of a great deal of scientific work, especially a great deal of medical research, which is of a 'fundamental' type, yet deliberately entered into to help create a baseline or framework from which applied work could grow. Such work the simultaneous Dainton Report had, more perceptively, identified and named 'strategic'.[4]

It is, however, still worth making a distinction between research seeking addition to knowledge and research seeking practical benefit, simply to clarify discussion, provided that it is understood that in practice, the distinction becomes hopelessly blurred. Practical research commonly produces major additions to knowledge, and fundamental research commonly yields important practical benefits. Nor do investigators restrict their objectives to one or the other, but can often change in mid-stream, as Dale did, from a rather practical study on ergot to what was (then) an academic study of the acetyl ester of choline. Strategic research naturally aims at both. There is the fact, too, that some (including myself) have not been able to identify *any* contribution to knowledge from which some practical benefit could not be envisaged.

If one is concerned over the justification of the use of animals for some set of experiments, it is possible sometimes to say that their purpose is primarily to increase our understanding and to remove ignorance about some bodily process, without really being able to say how it might be useful in other than the most general terms. At other times there is a more distinct practical goal, sometimes very specific as in testing toxicity or in choosing between possible new drugs or in following the metabolic fate of a substance in the body. In general we have a continuous spectrum, with each component in varying proportions. But the distinction has some value in that the warranty for the use of animals is probably seen as different in the two cases. Later, therefore, we shall distinguish between benefit by addition to knowledge on the one hand and practical benefit on the other.

## Summary

1. Experimental work is concerned with *what is not yet known*. This immediately implies that the outcome of such work cannot be specified in advance.

2. New knowledge was won over the centuries by reflection and by observation. But the great discovery, termed the 'way of experiment', around the seventeenth century, was of what could be gained by deliberate *intervention*, choosing test material and test procedures so as to simplify and disentangle various processes. It is the adding of deliberate analytic experiment to the observation of 'nature's experiments' that characterizes the experimental approach.

3. The outcome of an experiment is commonly not merely unknown but also *unexpected*.

4. The distinction between 'fundamental' and 'applied' research cuts across the grain of actual scientific practice: the term 'strategic' research more accurately describes that large body of medical research work which deliberately seeks to do work of a 'fundamental' type that will provide a framework within which 'applied' work may grow. But although the distinction is blurred in practice, it is worth keeping in mind to clarify the discussion of the benefits of experimental work, as flowing from new knowledge or new use.

# 4

# The ethical questions

Is it right in principle to do to an animal what you would not
do to a man?

Is it right to do to a man what you could do to an animal?

Is it right to do to a plant what you would not do to an
animal?

Is it right to do experiments on higher animals which could
be done on gravely handicapped humans?

Does not an animal have a 'right' to its natural life?

How can an animal have a 'right' if it has no responsibility?

The purpose of this chapter is primarily to consider certain arguments of principle. If strongly pressed, they can lead to the
absolute antivivisectionist position—that experiments on animals
should be absolutely prohibited. If less strongly pressed, the issue
ceases to be absolute, and the question of some sort of balance
arises. This involves detailed scrutiny of the gains and losses to
animals and man resulting from animal experiment and some
decision as to how that scrutiny should be conducted.

The reader is asked to review the questions at the head of the
chapter for a moment. They are a small sample of those that can
be asked: they show, incidentally, that the way a question is put
can place the onus of proof in one or other direction. The answers
depend, in one way or another, on one's knowledge of, and view
about, the relationship between man, animals, and the rest of
creation. The *general* issue now is whether or not there are differences between men, animals, and plants which justify different
treatment. Animal experiment may bring great benefits to man and
animals in the future, but they may also involve costs of pain,
suffering, loss of life, or interference with 'normal' life. So we
must establish the cost of the treatment as well as the benefit.

# The physiology of pain

The question of pain is so important for the discussion that it merits a preliminary brief account of some of our physiological knowledge about it. It is sometimes regarded as invariably and inescapably evil. But this is not true. Everyday life shows how it provides a valuable warning signal: grit on the eyeball, a thorn in the finger, chest pain with a coronary thrombosis, abdominal pain with peritonitis, the pain of a fracture, headache—the list is a familiar one. The signal is evidently of the risk of bodily damage, even if sometimes the signal seems inappropriate.

The physiology agrees with this[1]:

(1) Painful stimuli override all others. The electrical signals evoked appear very widely in the brain and not, like other modalities such as touch or vision, only in select areas.

(2) Accommodation to pain does not occur in the way it occurs, for instance, with touch, both as regards the nerve signals and the actual perception. This is illustrated by the way you soon cease to feel the spectacles on your nose or the watch on your wrist.

(3) If anything, the sense of pain 'recruits'—that is, gets worse with time; minor pains that seem quite tolerable at first can become very objectionable.

(4) There is a particular urgency associated with the sensation.

(5) The reflex physical response to a painful stimulus is very stable. Thus it is one of the last to be totally abolished under anaesthesia. The 'corneal reflex' (a blink on stimulus of the cornea) is a useful test of deep anaesthesia.

(6) The conditions in the tissues that give rise to signals in the nerves concerned are those where damage is impending but still reversible. Thus with heat, pain in the skin is felt when signs of cell damage are just detectable and when the temperature has reached a level that begins to damage enzymes.

(7) The pathway to the brain carrying the pain signal has many steps at which the final perception can be modified in various ways. These include other concurrent stimuli which can alter transmission upwards at the spinal level, as well as higher influences such as training, expectation, and attention. A striking example is the way a hack on the shin on the playing field may

be hardly noticed, although the same blow off the field would be found extremely painful.

The whole impact therefore is of something that must be attended to as soon as circumstances allow. Clinical medicine shows some reasons why. Perhaps the most vivid example I know is from a visit to the leprosy rehabilitation clinic at Vellore in India. Here one could see young Indian lepers with hands and feet that lacked pain sensation. As a result, casual damage, rat bites, and neglected infection had reduced many of their fingers and toes to stumps. One of the achievements of the clinic, under Dr Paul Brand, was to find ways of preventing further damage and enabling them to undertake work. An essential step was to train them to use their eyes to replace the missing pain sense, so as to avoid or detect damage to their hands and feet. For instance, in learning carpentry, they would be taught to look at their knuckles, and never to hold a tool so tightly that their knuckles blanched, which meant blood-supply cut off from the hand.

Another example is the rare condition of congenital insensitivity to pain. In a recent case[2] a 2-year-old child had an ankle fracture, which it exercised so much as to break the plaster put on it without apparent distress, and then went on to fractures of both arms and frequent skin injuries. She had a sense of light, limb movement, vibration, temperature, and pinprick (without finding it painful). It was remarkable that electric shocks to a nerve at mains voltage produced neither reflex withdrawal nor any complaint.

More familiar to the reader may be chewing the inside of the lips after local anaesthesia for a dental operation. A final example is how medical students have to be taught not to give morphine to relieve pain in an abdominal case until a diagnosis has been made, for fear of removing the critical localizing signs.

Considering all this, it is interesting that the main research effort is in finding ways of removing pain. It is as though, at an army headquarters, when a messenger arrives to report that a 'military' salient is threatened, the main effort is to gag him! A word should be said about the opioid substances that occur naturally in our own bodies: the enkephalins and the endorphins. It was hoped that these would throw a flood of light on pain and its handling. But this has not happened; nor have significant new analgesics resulted, despite intensive work. A major problem is that the layout of the pathways and their transmitter systems is far from established, and there is now a multiplicity of other peptides also claiming functional roles. At one time, it seemed possible that the occurrence of the opioids or their receptors might be a sort of marker of capacity for pain sense. But now they present as one among many, largely modulatory, transmitter substances, also occurring, for instance, in the goldfish retina, the snail, the slug, and the leech.[3]

Many, many people have pondered over the 'problem of pain'—how could a world be created that allows it? But on reflection, one can see that any system whatever that allows organisms to interact with each other or with the environment must have some signalling system to indicate any adverse result from the interactions. Not only gross physical damage, but also any interference with the actions of the organism must carry warning signals, whose meaning must correspond to what in us we call pain or suffering. As one sort of signal perceived as 'pain' diminishes in frequency, the next most severe takes its place. Indeed, the law courts show vividly how what counts as suffering today does not consist simply of pain and discomfort, but includes, for instance, the psychological pain of frustration or constraint. Thus, the only way of avoiding anything that could be classed as what we would call pain and suffering is a world either of inaction or lacking any freedom of interaction.

This does not, of course, remove the problem of inappropriate pain—though it is enlightening to try to design a system which would prevent it. Nor does it justify unnecessary pain. But it is a mistake to regard pain as an unmitigated evil; it is an inevitable accompaniment of interacting organisms. The essential task is to prevent *unnecessary* suffering, whether now or in the future.

The discussion brings out some points to carry forward. First, deprivation of pain sense—for example, local anaesthesia—can readily lead to self-damage. It is not a procedure to be used thoughtlessly. Second, what counts as pain varies with circumstance. Sensory decision theory is discussed later. Third, it is sometimes said that pain can never be measured. This is misleading. Sir Thomas Lewis's famous dictum that 'Pain cannot be defined, but is known by experience and illustrated by example' is a very wise one.[4] It will be argued later that by attention to human or animal behaviour, pains can at least be graded; and this may often be all that is needed.

## The changing issues

It is obvious, as one reads the older antivivisection literature, that pain takes pride of place. The views expressed were exceedingly trenchant and often absolute, calling for abolition, not mere control. As we search for the reasons behind them, we might expect an absolute objection to any suffering whatever; but in fact we often find acceptance of many veterinary practices of the day and of a good deal of temporary animal suffering for what were regarded as good reasons. The real motive force is revealed as horror at the contemporary or past accounts of scientists' experimental practices.[5] It was not that *no* suffering was the ideal, but rather that *this* particular suffering was intolerable, even if it was to advance knowledge. The idea of a balance between suffering now and preventable suffering in the future was implicitly recognized, for the arguments always followed two lines: on the one hand, the expounding of experimental horrors; on the other, a depreciation or refutation of the knowledge or prospective benefits to be gained, which is only relevant if some balance is to be struck. Some of the cases cited are indeed horrifying to a modern mind. But judging the past by our present standards is a misleading activity, and we need, before making a judgement as to what was or was not justified, to review the suffering at that time in homes or hospitals, in asylums, on the battlefield, in the factories, and in the countryside. When human suffering and loss of life were so common, animal life would hardly be valued any higher. This is not a mere debating point. Those doubting what life was like in the days before anaesthesia should read Fanny Burney's account of the removal in 1811 of one of her breasts for cancer. It was of about one year's duration, and the tumour was the size of a fist. The fact that it had invaded the underlying ribs, and so had to be scraped away, was just one element in 20 minutes of continuous agony.[6]

Citations of long-past examples of experimental work still appear in the literature, even though present circumstances make them irrelevant apart from their historical role in shaping institutions and traditions. The modern reader should be on guard lest information from that past is being paraded as current practice, just as he or she needs to be wary of work in other countries being presented as typical of this country. This happens quite freely, and references enabling one to check the allegations are usually absent.[7] The reader should ignore any stories of this kind, unless the citation includes date, country, and documentary reference sufficient to be able to order from a library. Some of the past literature, particularly the anti-vaccination propaganda, makes sad reading to a generation that has at last seen the abolition of smallpox. That could have happened earlier. Equally, there is a deplorable lack of explanation of the reasons behind the experiments. A protest by a distinguished group of scientists expressed very accurately the way much research has been described: 'The description distorted, the real reasons suppressed, the work trivialised, and motives questioned.'[8]

It is probable that the introduction of anaesthetics and improvements in experimental practice and the techniques available led to subsequent change in attitudes, since many experiments now clearly involved very little or no pain. A more absolute argument developed, the only *absolute* one I have encountered: namely, that it is absolutely wrong to inflict suffering or loss of life on any living creature. 'Suffering' is a more general term than 'pain', and can include anything from distress to death, and 'living creature' may include a narrow or a wide range of animals. An abolitionist position may be held while fully recognizing the cruelty of the natural world, its predators and victims, natural disasters, diseases, and the ordinary accidents of life; for it can be argued that, at the very least, one should keep one's hands clean by not adding to it in any way, so far as one can avoid it. This approach is accompanied, of course, by vegetarianism and other logical consequences. It is clearly a tenable attitude, if the vegetable kingdom is not included in that part of the living world whose existence or well-being is given absolute value. It may also be defended on economic, ecological, or health grounds, and may accompany a belief in herbal remedies. The price to be paid, of course, by this avoidance of particular sins of *commission* is (to other eyes) the growth of sins of *omission* and the rejection of all the benefits to animal and human life that have flowed, and can flow, from knowledge gained about the animal body, which will be described in Chapter 5. It would allow people or animals to die of preventable disease, just as, according to certain religious beliefs, it is right to allow a mother to die of haemorrhage after childbirth rather than give a blood transfusion. Other beliefs may support this position, such as the belief that it is good to abstain from action and to escape from the world of striving. That would entail a stepping aside from the living world as we see it, in which, at all levels, activity, effort, struggle, and seeking the limits of endurance are an inherent part. Total abolition also cuts off a large field of human biological enquiry, and accepts the corresponding permanent ignorance.

We may ask *why* should it be felt absolutely wrong for a man to kill or inflict suffering of any sort on any living creature? It seems likely that it is simply one of man's responses to the suffering around him because of his capacity to project his emotions on to other beings, to 'sympathize'. If we think of the conditions of human life at the time of the birth of Buddhism and the 'First Noble Truth' that all life is bound up with suffering, we can respect this impulse. But one major point arises. How does it come about that a man may be *blamed* for taking animal life, when we do not *blame* a cat for killing a mouse or a dog for worrying a sheep (however much we blame the owner)? We meet again a difference between men and animals, that it is only humans that are governed by words like 'right', 'wrong', 'ought', 'blame', 'responsibility'—words which express moral concepts. How it comes about that humans have a moral sense is a wider issue. But every time such terms are used, the difference between man and animals is reaffirmed.

## Consent

A second, slightly different argument of an absolute nature is that it is wrong to do experiments on animals 'because they cannot give their consent'. The discussion in Chapter 7 on the assessment of pain and suffering outlines the various ways in which evidence about these can be obtained in animals, and it is clear that animals can indicate suffering. But whereas it is possible to know whether a man, after due explanation, is willing or unwilling deliberately to override his normal avoidance of pain or suffering for some good cause, we cannot know this about an animal. From this comes the sense that it is 'unfair' to inflict deliberate suffering on an animal.

It is unfairness rather than suffering that is at issue. That very statement, however, accepts a radical difference between men and animals; both the conception of 'fairness' and the acceptance of it as carrying some moral obligation appear to be specifically human. If this is the case, the implication appears to be, not that it makes animal experiment impermissible, but that it reminds man again of the nature of his moral responsibility in respect of animals. This principle has in fact long been recognized. The revulsion at the use of curare (which paralyses voluntary movement) as an 'anaesthetic', that led to a ban on its use in this way in the 1876 legislation, was precisely because it deprived the animal of expressing reaction to its experience without preventing it from feeling pain.

The concept of 'consent' in an animal offers considerable difficulty. We do not know if animals consent to be domesticated, become pets, and be washed and trimmed, ridden, raced, treated medically or surgically, be exhibited, or be killed for food or as strays or for other reasons. Perhaps the underlying assumption of all legislation is that if any pain or suffering may be involved, then lack of consent is assumed. But it is hard to see much difference between saying 'We must minimize pain and suffering in animals because we assume they do not consent to it' and the simpler statement 'We must minimize pain and suffering in animals'.

We note, too, some more specifically human words: 'consent', 'fairness', and 'explanation'.

## Jeremy Bentham on suffering and killing

The absolute position becomes difficult when one has to choose between two courses of action both of which involve suffering. Inaction may be proposed as an escape, but only too often that too entails suffering. What to do, for instance, when a dog with a broken leg is found on the side of a motorway? It may be said that one must not base ethical systems on such difficult decisions, just as hard cases are said to make bad law. But an ethics which provides no guidance when choice is difficult is not ethics at all—just windy aspiration; who needs it when choice is easy?

So not many affirm the absolute position. Suffering remains an essential factor in the next stage of the argument. It is expressed in a famous footnote from Jeremy Bentham, the founder of utilitarianism, written in 1789 in a chapter on the limits of penal law:

The day *may* come when the rest of the animal creation may acquire those rights which never could have been withholden from them but by the hand of tyranny. The French have already discovered that the blackness of the skin is no reason why a human being should be abandoned without redress to the caprice of a tormentor. It may one day come to be recognised that the number of legs, the villosity of the skin, or the termination of the os sacrum are reasons equally insufficient for abandoning a sensitive being to the same fate. What else is it that should trace the insuperable line? Is it the faculty of reason, or perhaps the faculty of discourse? But a full-grown horse or dog is beyond comparison more rational, as well as a more conversible animal, than an infant of a day or a week or even a month, old. But suppose they were otherwise, what would it avail? The question is not Can they *reason*? nor *Can they talk*? but, *Can they suffer*?[9]

It is an important passage. Rather than stressing the difference between man and animal, with the resulting duty of care for man, it opens a comparability between them. Taking suffering as the essential criterion to be used, it introduces the idea of comparing adult animals with infant humans, the idea of animal rights (although elsewhere Bentham calls rights 'fictitious entities' and finds them confused), and the idea of animal liberation as comparable to the release of the Negro from slavery.

With any such quotation we need to distinguish between the argument in itself and the authority conferred on it by the name of the author. (The quoting of great names is a double-edged pursuit. One finds likewise that while one may gain inspiration from J. S. Mill's plea for personal liberty, yet one also finds him presiding at the East India Company at the peak of the opium trade with China, and a vehement opponent of state education.) It is therefore worth also citing what Bentham says a little earlier in the same footnote:

There is very good reason why we should be suffered to eat such of them as we like to eat, we are the better for it, and they are never the worse. They have none of those long-protracted anticipations of future misery which we have. The death they suffer in our hands commonly is, and always may be, a speedier, and by that means a less painful one, than that which would await them in the inevitable course of nature.[10]

With the Benthamite move, then, we abandon the absolutist position for an intermediate one: humans are entitled to kill animals—for example, for food—because the suffering involved is so slight, especially compared to that involved in their natural deaths; but animals' capacity to suffer is now claimed to give them rights against more serious suffering, comparable to human rights against racial oppression.

## Sentiency, purposiveness, self-enrichment, experiential life

To analyse this new position, let us first focus on the specific suggestion that Singer, following Bentham, has urged most strongly.[11] This is that the essential criterion is 'sentiency', a convenient shorthand term for the 'capacity to suffer or to experience enjoyment'. It is this that gives rise to 'interests', comparable to human interests, of which we must take account. Whenever 'sentiency' is present, there is an 'interest' which must be weighed equally with other interests, such as human welfare or that of other animals, knowledge, and the like. The argument is not for equal treatment but for equal consideration. (There is some obscurity and scope for arbitrariness here; if the treatments are in the end different, how do we know that equal consideration was in fact given? By testimony under oath?) This leads to a classical utilitarianism, in which the consequences of any action are to be calculated in terms of the overall good to be derived, as against the overall suffering entailed. One difficulty is that we cannot know what the animals would prefer, although with humans we can ask for their preference; the human has to speak for the animal. In a general sense, this sort of approach would be used by many, and the real difficulties arise in the weight to be ascribed to any 'benefit' in the measurement of 'suffering' and in balancing the resulting incommensurables.

But difficulties remain. Thus, any subject, human or animal, treated effectively with pain-killers or suffering from the rare congenital freedom from pain (mentioned earlier) would have to be regarded as non-sentient, and therefore eligible for experiment. This has prompted an alternative proposal: that it is because 'animals and humans are so organized as to be purposive creatures with various desires, drives, intentions and aspirations' that animals acquire 'rights'—to be left alone, to pursue ends, to be allowed to live.[12] The extent of interference with 'purposiveness' becomes the key. A further variant arises in a slightly different way, from a search for some defining quality most likely to do justice to any fundamental difference between human and animal life; this is then found in the 'capacity for self-enrichment', as exemplified by things which make life valuable:

the pleasures of friendship, eating and drinking, listening to music, participating in sports, obtaining satisfaction through our job, reading, enjoying a beautiful summer's day, getting married and sharing experiences with someone, sex, watching and helping our children grow up, solving quite difficult practical and intellectual problems in pursuit of some goal we highly prize, and so on.[13]

Yet another criterion has been advanced, by T. Regan, for assessing the relation between man and other creatures: that of 'inherent value'.[14] The idea that the possession of life itself has inherent value—that is, Schweitzer's approach—is dismissed, because it would assign such value to 'crabgrass, lice, bacteria and cancer cells'. Instead, inherent value is restricted to those who are the 'subjects of a life' or who have an 'individual experiential welfare' or who have a 'psychophysical identity over time'. In practice, it is proposed to restrict it to *Homo sapiens* aged 1 year or more (provided they are not 'very profoundly mentally retarded') and to mentally sound mammals of 1 year or more.[15] In Regan's hands this position seems at first to lead to an absolutist position: vegetarianism, dissolution of the animal industry, separation from wildlife, abolition of hunting and trapping, abolition of toxicity tests and of use of animals in research, and abolition of any distinction between animals on grounds of commonness or rarity. But when it comes to conflict of interest, this proves not to be the case. Regan writes, recalling the 'lifeboat' case:

Recall the situation: there are five survivors, four normal adult human beings and a dog. The boat will support only four. Which one should be cast overboard? The rights view's answer is: the dog. The magnitude of the harm that death is, it has been argued, is a function of the number and variety of opportunities for satisfaction it forecloses for a given individual, and it is not speciesist to claim that the death of any of these humans would be a prima facie greater harm in their case than the harm death would be in the case of the dog. Indeed, numbers make no difference in this case. A million dogs ought to be cast overboard if that is necessary to save the four normal humans, the aggregate of the lesser harms of the individual animals harming no one in a way that is prima facie comparable to the harm death would be to any of these humans. But suppose, a critic may conjecture, it is not a question of having enough room on the boat. Imagine it is a question of which individual to eat if four others are to survive. The rights' answer, once again, is: the dog.[16]

We may notice, in passing, that the criteria proposed progressively introduce an element involving others. Pain is intensely individual, while purposiveness, self-enrichment, and full experience carry strong social elements.

We have, then, sentience, purposiveness, capacity for self-enrichment, and psychophysical identity named as fair tests for examining the comparability of men and animals. They are put forward as the soundest ways of testing for a real distinction between human and animal life. They all rest on a search for some specifically human quality, and assume that it can be articulated in these sorts of terms. But all the authors concerned conclude that no clear distinction can be seen.

At this point two counter-arguments may be put, maintaining the existence of such a distinction. First, a reviewer of Singer's book *Animal Liberation*, D. L. Hull, has noticed that Singer accepts evolutionary theory and uses it to support his case in saying that only a 'religious fanatic' can continue to maintain that *Homo sapiens* is different and distinct from other species. But Hull points out a potential fallacy in trying to rank species by their 'likenesses'.

Species are grouped into higher taxa because of descent, not degrees of similarity. From the biological point of view, the relations that exist between races and between sexes of the same species are different in kind from those between species. If the principles of evolutionary theory are to be taken seriously, there are excellent reasons for us to exhibit a greater moral commitment to a child than to a porpoise.[17]

Second, and more important, is the fact that, in animals, capacity for self-enrichment and the other qualities are a pale caricature of the human quality. We have already, in the first chapter, seen that although one may recognize a continuity in the scale of creation, nevertheless there is an operational 'upper bound' to the animal world between man and animal, based on his capacity to accumulate and share his experience by language, by the spoken and (especially) the written or printed word. It is this—humans interacting with other humans and building on other humans' achievement—that has opened up a difference so great, and still growing, as to have created a qualitative distinction in nature and value between the human and animal worlds.

## The higher animal vs. the 'lower' human 'incompetent'

An important step in the argument comes with each of these approaches, when animals are compared, not with normal humans, but with what has been termed the human 'incompetent'. Specifically, it is claimed that some higher animals (typically, a monkey is cited[18]) would suffer a greater cost from experiment than some humans (infants, the senile, the seriously diseased, or the mentally handicapped), because of the animal's superior capacities compared with these humans. The dividing line separating man and animals is then supposed to be destroyed.

This comparison is made because a direct equation of *normal* man and *normal* animal is ultimately unconvincing. But human compassion for the infant, the old, the handicapped, and the diseased human (even if such compassion is not exercised by animals) provides an opening. It is argued that if man treats these human 'incompetents' as normals, he should treat animals of higher capacity at least as well. It is not entirely clear whether those taking this position are serious in suggesting that experiments presently done on animals should actually be done on humans. On the whole it seems probable they are not, but rather that these comparisons are a dialectical device that exploits human compassion as it tries to suggest that the experimenter should do experiments on incompetents if he is to be consistent. It should be noted that in human experience of illness and suffering, some argue that the deepest human qualities are revealed.[19]

This comparison between the higher animals and 'lower' humans deserves close attention. As already said, it represents an attempt to blur irretrievably the difference between man and animal, but (interestingly) it draws back from any similar attempt to blur the differences between animals and, say, vegetation. A number of comments may be made:

(a) The method is immediately open to a question of logic. It is illogical to compare the normal adult of one ostensibly 'higher' species with an infantile, senile, defective, or diseased sample of another. As an analogy, it would be similarly wrong to make a comparison of the most intelligent woman (or man) with the most stupid man (or woman) as a reasonable way to compare the capacities of the two sexes.

(b) In choosing between groups where additional suffering or deficit might be involved, it is inequitable, as well as inhumane, to regard as eligible those already suffering or handicapped.

(c) Some of the cases cited of 'lower' humanity seriously underestimate their capacities. Thus a patient with spina bifida is used as an example. Yet spina bifida may be symptomless, or damage may be restricted to loss of bladder or anal control, or there may be some neurological deficit, mostly to the lower limbs. It is essentially a spinal disease, leaving the brain untouched, and is compatible with a full range of human activity. If such cases are to be used, then severe cases of poliomyelitis (e.g. Franklin Roosevelt) are equated with animals too. Down's syndrome is regarded as a sufficiently serious human handicap for some to have the individual killed *in utero* by abortion. Yet those familiar with individuals suffering from it or with comparable cases know that they can learn, teach others, earn money, and join in family and community life as equals; their experience today, from their own efforts joined to those of normal humans, points the way to still further development of their capacities. Similar remarks about the senile and the diseased are in order. As for babies, while Bentham may not have found them 'conversible', the astonishing growth in organization of the nervous system and in mental capacity taking place in them from birth is as remarkable as anything that was going on in his own brain.[20]

If these cases have to be abandoned as suitable examples of overlap, it may be said that the postulated 'incompetent' is still more deficient, 'very, very, severely handicapped'.[21] But as the 'very's' multiply, the difference between normal animals and normal man is emphasized. If you have to suppose some human dragged down with such handicap as to lack all human interaction, to be without hope of recovery, and yet to be still counted as living, credibility is stretched, and the proposed bridge between animal and man becomes instead a further illustration of the difference.

(d) The human 'incompetent' cases have, of course, all been chosen for their handicap. In the past, such a case might have been a cretin (someone suffering from mental retardation due to thyroid insufficiency). Suppose one had been chosen for experiment in 1890. A year or two later, cretinism was preventable and curable if treated early enough, and it would have turned out that someone with full human potential would have been used. Phenylketonuria, a biochemical disorder, provides a more recent example. The baby grows up. The new molecular biology means that the seriously diseased cancer patient today need not *totally* despair of cure. The possibility of transplant of nervous tissue to replace damaged areas of the brain is already achieving its first successes, and biochemical approaches to understanding senile dementia are very promising. The word 'hopeless' is sometimes used to describe such cases; yet just because it is defective *human beings* that are being postulated, an 'up-side', a scope for betterment is immediately created. We should assert that strictly there are no 'hopelessly' defective human incompetents—all are potentially competent.

(e) Finally, as to practice, one cannot regard as scientifically satisfactory a proposal that, regardless of what experiment is to be done, it should be done not on a normal organism, but on one that is infantile, senile, defective, or diseased.

The reader may well feel at this point that the obvious has been laboured sufficiently, that the exploitation of the concept of the human 'incompetent' is both insensitive and ineffective, and that he does not need convincing that it is better to use even a chimpanzee than a human if an experiment is of such importance that it must be done on one or the other. Singer evidently feels the same; but he, too, labours the point because his belief, that there is no argument *in principle* for preferring experiments on animals to those on man, underlies his doctrine of 'speciesism' with all its widespread implications.

That doctrine—namely, that to discriminate between human and animal species is as wrong as discriminating between races or sexes—rests on selecting likenesses and ignoring differences between man and animals.[22] Without repeating what has been already said, a short answer is: 'If you are unable to articulate the difference, then just look around you at the respective achievements.' But there may be a lesson, in the dangers of jargon. The term 'speciesism' is quite vacuous. It merely says that humans and animals share one principle of classification: namely, being of various species. They are also all made of carbon, hydrogen, oxygen, and other chemical elements; so that distinctions between their chemical constitutions should no doubt be termed 'elementism'.

It may be felt extraordinary that it should ever have been contended as a general proposition that the relationship, for example, between a man and his wife or a colleague of another race should be seriously equated with his relationship to a chimpanzee or other animal. Man *has* travelled further than animals along an evolutionary pathway. It *is* possible to draw an unequivocal biological and objective distinction. But it has been worth labouring the point for another reason. One purpose of the animal welfare movement used to be to sharpen our sympathy for animals. It is hoped that this discussion of the 'human incompetent' will help to sharpen our sympathy, too, for the infant, the senile, the diseased, and the handicapped.

## Superbeings from another planet

As a brief digression, one may mention a line of argument, in a sense complementary to that about the human incompetent, that is resorted to from time to time.[23] I first met it when I was sent a little bookmarker by an antivivisectionist group. It showed a Martian figure, knobby and armoured, holding a writhing scientist upside down over a cauldron of boiling oil. In this and similar writing, the animal-user is being asked to imagine himself being treated by a higher being as he (reputedly) treats animals. As Colin Dollery once remarked, 'I will take the matter up with the Martians when I meet them.' More seriously, one has to note an intrinsic frivolity in these arguments. 'Super-beings' may be postulated, but the same old mental furniture is unpacked in their fictitious utterances, with nothing 'higher' about them.

There is a genuine interest in considering what a next step might be in a progression such as stone → grass → animal → human → higher being—although one must remember that 'progress' is a suspect word to some. It would require a difference between human and higher being as great as that between animal and man. This is hard to imagine, and E. A. Abbott's *Flatland* is a fascinating illustration of what is really involved in such thinking. He was trying (in 1875) to make vivid the envisaging of a fourth spatial dimension, by portraying what happens in imagined transitions between life in zero, one, two, and three dimensions.[24] For example, to a two-dimensional being, a three-dimensional visitor could be seen only in a continuously changing cross-section, with a miraculous capacity to go through solid walls. That illustrates what is at issue. One is not considering a trivial dressing up of familiar human behaviour, nor an equally trivial enlargement of computer size. A serious step comparable to that from animal to man would have to be as great as that from a bird's nest to Salisbury Cathedral or from a cat's mew to Shakespeare. Music and poetry certainly express such strivings—but that is not what the super-beings from another planet in their science fiction 'gear' are concerned with.

## Animal 'rights' and moral worth

None of this discussion, of course, ought to weaken our projective sympathy for animals. But it leaves undiscussed the question of animal 'rights'. We can readily accept that sentience, purposiveness, capacity for self-enrichment, psychophysical experience, as well as animals' more familiar characteristics, give a 'worth' to their possessor. In turn, that worth implies a 'duty' of care among humans, by virtue of that faculty (not shared by animals) by which we accept such a duty. If we accept that duty, then we acknowledge a moral worth. It can then be argued that this confers a 'right' on the animal. But it does not necessarily follow that because something has a worth it therefore has rights. We may feel a worth in some beautiful object (such as a building, a picture, a piece of porcelain) and a 'duty' to look after it, but we would not admit that it had a 'right' against us. If we are looking for a convenient terminology, we could say (following Linzey) that we have a duty not 'to' the object but 'in respect of' it.[25] But once we do this, we break with the idea that duties necessarily imply 'rights'; so it would seem that the existence of rights must be decided in some other way.

Indeed, as we struggle with these ideas, which are the subject of a profuse and conflicting literature, the idea of 'rights' seems to become more and more confusing and less and less useful save for political purposes. In a classic, now nearly 100 years old, Ritchie has argued that the idea of rights has originated in three ways.[26] To begin with, in an earlier, more despotic world, rights were conferred on subjects by whoever was in the position of ruler; they could almost be described as 'allowances' of what the subjects could have and do. Then the feeling that this was arbitrary and inequitable generated a claim for 'natural rights', based on the claim that the mere fact of existence created a series of claims— for instance, to freedom, continued life, food, shelter, justice, education, and so on. But then the questions arise: What is their justification? What if claims conflict? How are just rights to be agreed? How are they to be met, and from whose resources? As Ritchie says:

In the chaos of conflicting individual impulses, instincts, desires and interests, we can find no stable criterion. We must go beyond them to the essential nature of things. But what part of the nature of things is here relevant? Is it not simply—human society?[27]

It seems that if 'rights' as a concept has any real value, then they are inescapably a function of society, of the interaction of human beings. We can abandon natural rights very simply and, following Caplan, distinguish 'moral agents' and 'moral objects' (other terms are possible).[28] The former are capable of moral choice and can accept duties; they can therefore make a claim to be accorded correlative rights by others, and may feel correlative duties for themselves. It seems to be agreed that the only moral agent of whom we have knowledge is man, although we can, not unreasonably, trace vestiges of the origins of his moral capacity in animals.[29] 'Moral objects' are those in respect of whom man has duties; but they themselves have no moral choice, have no duties, cannot make claims, cannot be responsible, and are not participant parts of society. There are thus many more moral objects than moral agents.

It is not always noticed that there needs to be some equivalence between 'rights' claimed and duties accepted. If rights are claimed and there is no one on whom the duty of satisfying them can reasonably be laid, then the claim is an empty one; these are not rights but aspirations. But provided that both duties and claims are restricted to moral agents—that is, to humans—the necessary 'balance of trade' between rights and duties is possible: the totality of claims justly made can be arranged to be covered by the totality of duties justly discharged. The duties of a moral agent, therefore, are to other moral agents, as well as in respect of moral objects. Human rights is thus a possible concept, though whether it is a useful one is another matter.

As regards animals, however, the term 'rights' is profoundly misleading. What is meant has little to do with human rights, which can be claimed, justified by a claimant, forgone by a claimant, matched to duties, revised, agreed, voted on by the beneficiaries. Animal rights are 'fictitious entities' (to use Bentham's term[30]) assigned to non-cooperating animals on the basis of arbitrarily selected criteria. The cause of animal welfare is not in the end assisted by basing it on arbitrary misuse of language. The concept of animal 'rights', with its false analogies to racial and sexual rights in humans, is replaced with no loss and perhaps some gain of clarity by terms such as the 'moral worth of animal life', if special terms are needed. But the old language of animal welfare would suffice.

## Some other approaches

The reader may have noticed that the discussion began with individual suffering, but gradually expanded beyond the individual to include others. Who is it that is now 'counting' the cost—the sufferer or someone else? It seems that, in general, it must be someone else: namely, the human (or humans) who is seeking to decide on the rightness of some action. A suffering animal or man, in isolation, is just that—isolated. It is only by sympathetic projection or communication that another being makes contact. That other being may 'speak for' the sufferer, but cannot 'be' them. How then is justice to be done to the individual? We could argue that one of the great pathways of ethical advance has been in this direction, in the sharpening of 'projective sympathy' (the word 'projective' may seem redundant, but 'sympathy can have a cosy self-wallowing flavour, and 'projective' stresses the mental movement outwards). So we ask not just 'What do I want or think?', but other less egocentric questions such as 'If everybody thought like this, what would be the result?' (a simple version of Kant's question) or 'If I place some other, arbitrarily chosen creature in my place so as to avoid all bias, using whatever anthropomorphic or human insights I can muster, what would be said?'. These are essentially social thoughts, implying discussion and agreement.

This brings us close to one particular ethical system, Rawls's theory of justice which centres on fairness.[31] An example would be an agreement between two children faced with the last piece of cake that it would be fair that one should divide it in two and the other choose which piece to take. Many find attractive the principles of (1) agreeing on a fair procedure and (2) securing that in that procedure an individual's part in establishing rules is kept separate from that individual's possible benefit or loss when the rules are implemented. It resembles the principle of 'indifference' used in economics and psychology, where things are set equivalent when in a series of choices one is chosen as frequently as the other. A notable point is that, unlike systems of rights, it admits from the start that fairness imposes constraints on one's own liberty as much as on that of others and that duties are called for from you to others, as well as from others to you. It is striking that in the ethics of 'rights', the words 'correlative duties' are often used only in respect of the duties imposed on others in meeting the rights you claim. It rarely refers to the duties to others that you should accept in exchange for the rights given to you. An interesting exercise is to look through the various declarations of rights over the decades, to see how rarely duties *by the claimant* are stated.[32]

On this approach, since an animal cannot assess justice, cannot enter into an agreement about fair procedures, cannot claim rights, and cannot undertake duties, it must be treated separately from the human. There is an analogy with the question of consent by an animal, discussed above. None of this, of course, alters human obligations with regard to animals; it simply helps to clarify their true basis.

A rather different, and famous, approach is that of Albert Schweitzer (see Appendix 3). His 'reverence for life' did not originate in a general benevolence, but as a solution to a problem in synthesizing two lines of ethical thought. His thinking led him to a questioning of personal rights without duties, an absolute rejection of violence, and a rule to preserve life and not to kill any living thing (down to bacteria) unless there is a necessity. He does not play down either life's conflicts or the moral conflict for man that results: 'One existence survives at the expense of another of which it yet knows nothing.' But evolution has enabled man to know of the existence of other 'wills-to-live'. So the conflict can have a sort of resolution, whenever man uses his life to preserve life, even at the humblest level. 'If I rescue an insect from a pool of water, then life has given itself for life, and the self-contradiction of the will-to-life has been resolved.'[33]

A third approach must not be forgotten—that of the Judaeo-Christian tradition. Although this lies outside a scientific discussion, it often appears in the background and is a major cultural element.[34] The concept of the 'dominion' of man over animals comes from it; but if one turns to the original literature, the picture is not quite so simple. This is outlined in Appendix 4, but the essential thoughts emerging from it may be summarized thus: from the Old Testament, a great enjoyment of animal life and a duty of trusteeship under a Creator; from the New Testament, the value of all creation, but especially that of man, exemplified not by the learned or the powerful, but by the child.

## The specifically human

Throughout this chapter and earlier, a number of specifically human capacities have been mentioned; and it is worth bringing them together briefly:

(1) the assigning and acceptance of blame, duty, and responsibility;

(2) the ability to consent;

(3) the concept of fairness and the capacity to make a contract;

(4) the counting of cost;

(5) projective sympathy and compassion for other beings;

(6) reverence;

(7) the concept of trusteeship, or stewardship;

(8) social feeling.

In a discussion on animal experimentation at the Royal Society in 1983, the comment was made that an answer to accusations of 'speciesism' is simply 'humanism'. The human being should respect all species, including his own, and he is as entitled as any to attend to his own welfare. A humanist approach goes further than this, as the capabilities above suggest, in that man can feel concern for species other than himself. For the first time, a being who can act on behalf of all other beings, has appeared. One fashionable view seems to be that the appearance of man on the planet is an ecological disaster[35] and that no good action can be expected from him (especially his scientists). Yet, if one envisages some impending world catastrophe—say a virus pandemic affecting all living creatures[36] or some other apocalyptic threat—it is only man that could have some hope of finding a solution for saving that living world. Even for those who prefer animals to man, it would seem prudent to give priority to man's survival.

## The ethical synthesis

> His habit was to encamp near to the region of practice in all his philosophical enquiries.
>
> W. E. Gladstone on Bishop Butler, author of
> *The Analogy of Religion*[37]

A great difficulty in ethical discussion lies in the lack of agreement that emerges for the layman as he reads the philosophers. It is as though men do not agree (in respect of words like 'pleasure', 'good', 'right', 'fair', or 'ought', and their opposites) as to which of them derives from which, if indeed any of them have such an origin and do not have another source, although it is no part of this book to analyse these differences. Technical terms also multiply quite freely, and delicate distinctions, that turn out to be not particularly material, are drawn. Some will find the stress on the 'human incompetent' unappealing, together with its revelation of some philosophers' images of the infant, the old, the ill, and the handicapped. A good deal of arbitrariness over 'cut-off points' lies concealed in arguments ostensibly from principle. There is a gross asymmetry in the attention paid to analysing animal and human capacities respectively, the latter being almost trivialized sometimes. A striking feature is how specifically human words and concepts multiply; as the attempt to equate man and animal proceeds, the language used demonstrates the difference. How to act when interests, rights, or duties conflict emerges as an important test for the various theories.

On the other hand, if, to use Gladstone's phrase, one encamps 'near to the region of practice', one finds that in the outcome there is a good deal of general agreement: for instance, that it is good, right, and fair to try to diminish human and animal suffering, and that one ought to do it.[38] When choices must be made, it seems that we have to balance matters up in our own minds and then balance them up again between each other. The various ethical approaches appear in the role not of prescriptive guides, but more of models, whose consideration can make the mind look at the issues from another standpoint and sensitize it to interests that may have been disregarded.

The benefits accruing from animal experimentation have not yet been considered, so the material for striking a balance is not yet to hand. One might note, however, how in the debate the strict utilitarian, with his 'felicific calculus' for achieving such a balance, is often ridiculed. Yet he might feel a little contempt for some of the remarks; for are we not doing much the same as he, balancing moral principle, knowledge, possible practical benefit, and suffering by a method wrapped in mystery, sometimes facilitated by 'intuition' but usually dignified by the name of 'judgement'? This final process seems in fact to be a sensible mix of 'do your best utilitarianism' (e.g. numbers *can* count) and 'deontology', the sense of 'ought'. We try to balance things as best we can—and so we ought; and it is on that balance that we think we should act. In the last chapter, we shall consider the relationship between humans and animals, which provides the context for our decisions.

## Summary

1. Opposition to animal experiment was initially not on principle, but on the grounds that the amount of animal suffering, as perceived by the critics, was too great to be justified by the knowledge gained. With the introduction of anaesthetics and other advances, the ground of debate has changed. The absolutist approach questions on principle the justifiability of inflicting any suffering or loss of life on any animal, for any reason at all. Use of animals may also be said to be unfair because, unlike man, they cannot consent. Absolutist arguments face great difficulties when there is a conflict of interests.

2. Commoner is the Benthamite argument that there is insufficient difference between animal and other species to justify

different treatment. In this 'comparability' approach, various criteria, such as 'sentiency', 'purposiveness', 'capacity for self-enrichment', and 'psychophysical identity', that might justify distinguishing man and animal have been considered, but found inadequate. These criteria, however, fail to represent human quality; and it is argued that it is the same characteristic of man that defined the 'upper bound' (pp. 12–15) which also gives human life a special value. This characteristic is the capacity to accumulate and share experience by spoken, written, and printed word, with all that flows from this, including the capacity to consent, to ask moral questions and feel moral obligations, and to feel responsible for animals. Its significance is measured by the range of human achievement compared to that of animals.

3. A particular test case, that of the 'human incompetent', has been proposed. It is suggested that it is as justifiable to do experiments on an infant, senile, defective, or diseased human as on a healthy 'higher animal', because of the latter's superior capacities. This is rejected because: (1) it is illogical to compare the defective of one group with the best of another; (2) it is inhumane to add to the disadvantages of the already disadvantaged; (3) the capacities of these human 'incompetents' have been seriously underestimated; (4) the human cases have been chosen on account of a handicap, but this immediately creates the possibility of their being restored to human normality; (5) it is doubtful if there will be any scientific requirement for experiments done exclusively on the infantile, senile, backward, or diseased.

4. More generally, it is indeed remarkable that a man's relationship with, say, his wife or with a colleague of another race should, as a general proposition, be equated with his relationship with an animal.

5. The case for animal 'rights' is reviewed. There seems no basis for a claim for natural rights for animals, since they can neither claim them nor discharge correlative duties; and there are arbitrary elements in its application. But one can readily distinguish between humans as moral 'agents' and animals (amongst others) as moral 'objects' in respect of which humans may accept duties. The term 'animal rights' should be replaced, with no loss, and perhaps some gain, by such a phrase as 'the moral worth of animal life', if the traditional language of animal welfare is not good enough.

6. Other approaches are based on Rawls's theory of justice, Albert Schweitzer's 'reverence for life', the biblical tradition, the set of specifically human capacities, 'humanism', and the fact that only man can act to protect the natural world. They all point simultaneously to a fundamental difference between man and animal and to man's responsibility. The cost−benefit balance between animal and human suffering now and ignorance and suffering to come can only be left to informed human moral judgement, both individual and collective.

# 5

# The benefits of animal experiment

In Chapter 3 we discussed the difference between 'fundamental' and 'applied' research, and concluded that in experimental practice research workers are constantly mixing these approaches or switching from one to the other. The labelling of a piece of work as one or the other is therefore often difficult. Nevertheless, in trying to assess the benefits won by animal experiment, it remains useful to distinguish the benefits of addition to knowledge as such and the benefit of some practical gain. The distinction corresponds to Francis Bacon's *experimenta lucifera*, experiments shedding light and dispelling ignorance, and *experimenta fructifera*, experiments yielding fruit. Often it has been experiments shedding light, work that illuminated our general understanding or created a sense of pattern and coherence in the world, by such as Newton and Einstein, that have been most admired; more practical work (on steam-engines, mackintoshes, or papermaking) has been appreciated but relegated to the position of a cart-horse compared to a Derby winner. In discussions about animal experiment, however, it has largely been the other way round. It is sometimes claimed that only practical benefits can justify the infliction of suffering on animals and that the motive of satisfying man's curiosity is not enough. 'Curiosity', as a word, like 'inquisitiveness', has perhaps now acquired a rather gossipy, nosy, trivial flavour; but we lack a better one for that exploring human faculty that wants to diminish ignorance and increase understanding.

To assess these two types of benefit, then, we shall use a historical method—namely, a 'test of deletion'—whereby we first seek to identify those things which we now know and can do which were made possible by animal experiment. Then we suppose them removed, deleted from our knowledge for lack of the necessary experiment; then we see what the loss would have been. Some of the arguments against animal experiment are absolute, and they would have been equally valid at any time in past centuries, even thousands of years ago; so the test of deletion may go back equally far. Other arguments against animal experiment are directed against particular features, such as psychological experiments, so a more restricted deletion is necessary. Sometimes the argument is not absolute, but simply wishes greatly to reduce animal experiment at once; one has then to attempt an estimate of the quantitative reduction of benefit, rather than its total loss.

It seems that in the end it is necessary to use a historical method of this sort. As already stressed, it is an essential characteristic of scientific research that it is concerned with what is at present unknown, just as is the product of any other creative activity. Elgar, walking perhaps on the Malvern Hills, said 'Music is in the air all around, you just take as much as you want.'[1] It is the same with scientific discovery: it is all there waiting to be found, if only you can do it. But you cannot assert with the force of logical certainty what will be discovered, or undertake to make a particular discovery. Consequently it cannot be asserted that in the future, if animal experiments were abolished or reduced, this or that particular thing would fail to come about. You can make some guesses but you cannot be sure. Bacon (see epigraph to Chapter 6) comments most perceptively how a discovery seems incredible before it has been made and a foregone conclusion after; and this psychological fact dominates much of the debate. It is a difficulty that the animal experimenter has faced since at least 1840, when Marshall Hall was challenged as to how he could guarantee that his experiments would indeed produce a 'good'. Like him, we can only look back and show the sort of things that would have been lost if experiments had been stopped in the past. As in other matters, our confidence in future benefit is a function of the historical record.

## Benefit from knowledge

We noted that sometimes new knowledge won by animal experiment is dismissed as 'merely' satisfying the curiosity of the experimenter. In a university, we take for granted the value of free enquiry and knowledge; and our task is to understand past knowledge, to interpret and preserve it, to teach it to the coming generation, and, by building on it, to seek to add to it. Although that very spirit of enquiry accepts that it is legitimate to question its own value, yet, as with any question about long-accepted values, it is not always easy to articulate the answer. We can start by emphasizing that it implies a removal of ignorance, and that the failure to win new knowledge means a perpetuation of ignorance (darkness, to continue Bacon's metaphor).

But let us apply our test of deletion, in a sort of 'think-experiment'. Suppose that no animal experiment had taken place for, say, 2,000 years, although scientific work on the inanimate world had continued unrestricted. Would the resulting ignorance and lack of understanding matter? People vary in the importance they attach to knowledge about ourselves and the world around us. But whether highly prized or not, we would now have a quite remarkably one-eyed view of the natural world. On the one hand, physics and chemistry and their precursor and daughter tech-nologies could be fully advanced. There seems to be no discovery or insight in these subjects for which animal experiment was necessary. Galileo, Kepler, Newton, Dalton, Lavoisier, Faraday, the Curies, Rutherford, Einstein, and the engineers, the industrial chemists, the miners, the explorers, astronomers, geologists, meteorologists—all these could still have gone forward. On the other hand, our knowledge of the functioning of ourselves and of the animal kingdom would be inferior to that of the ancient Greeks. To understand what that would mean, let us start with an experiment by the famous Greek physician Galen (AD 129–c.200), whom we can take as possibly the first significant animal experimenter.[2] It is extraordinary today to realize that one of his experiments was to show that the arteries contained not air but blood. He did it by tying an artery in two places and making an opening in between. How is it possible that this was not always obvious? The answer is probably that after death the arteries contract and empty themselves, whereas the veins remain full of blood. It would be rare to *see* an artery in life, without deliberate vivisection, since it retracts into a wound. That there was a pulse was well known, but what was pulsing, and in what direction, was another matter. It was not until the thirteenth century that it was shown that you could recognize arterial bleeding by the way it spurted. Indeed, blood itself was thought of quite differently: it was obviously one of the body fluids, but that it circulated was unknown (the very idea of a pump lay far ahead), as was the need of that continual circulation for the health of the whole body. It would not have been clear at the time that loss of blood was especially undesirable, any more than loss of urine or stools or vomit; in fact, to think of some undesirable humour being carried away in blood-letting makes some sense of that practice. So it is not surprising either that the tourniquet to stop blood loss also lay far away in the future.

The anatomy of the heart was, of course, long known in a very rough way. But it is remarkable that a treatise of the Hippocratic corpus on the heart, written in about 280 BC, that quite correctly describes the valves of the heart at the root of what we would now call the pulmonary artery and the aorta, and their function in making it impossible to blow air or pour water into the chambers of the heart through these vessels, adds 'especially on the left, for that side has been constructed more precisely, as it should be, since the intelligence of man lies in the left cavity'. Galen helped to counter this, in demonstrating the control of the body by the brain, by discovering that in a live animal, cutting a particular nerve in the neck known to come from the lower brain abolished movement of the larynx; this showed a control of speech by the brain, as well as making clear that nerves in some way controlled muscles. In a dead body, nerves and tendons look quite alike; both the Greek word 'neuron' and the Latin 'nervus' can mean either nerve or tendon. Only with experiment in life could the distinction become obvious between a tendon, essentially a tough cord transmitting mechanical force, and a nerve, which carries an activity that causes muscles to contract in response to reflex or volition. So we would not have known that the pinky-grey pulp in our skulls was what we now call a central nervous system, the 'seat' of intelligence, consciousness, learning, memory, control of voluntary action, mood, and emotion. Our thoughts, our vocabulary, and our outlook on the world, on ourselves, and on animals would be profoundly different. The very knowledge of our relationships to the animal kingdom and of the evolutionary connection and our understanding of their needs and behaviour that in part motivates the modern animal welfare movement would be lacking. If we reflect on how unbalanced our views would be, we can say, indeed, that we *dare* not cease trying to understand how our bodies and those of animals function: we dare not let our knowledge of the inanimate world outstrip our knowledge of its inhabitants.

## Practical benefits: and some caveats

This is a huge subject. Figure 1 provides a summary picture of part of it. The growth of animal experiment is shown here since 1890, using a logarithmic scale, so that it describes the proportionate increase in successive decades. Beneath it are given the names of particular substances or drugs, or sometimes classes of drugs, that have been important in human or veterinary medicine and that required animal experiment for their introduction. The approximate date of their introduction is opposite the name concerned, so that we can see the successive 'ages' of the vaccine, the vitamins, the antibiotic, and some other contributors to the therapeutic revolution.

Some caution is needed before making claims about medical progress, since a number of criticisms may be launched.

First, the progress may be regarded as grossly exaggerated. Thus some remarks by Dr David Owen, a one-time Labour Minister of Health, have been claimed as grounds for scepticism: 'A man aged fifty in 1841, when reliable records began, could expect to live a further twenty years; by 1972–4 a man aged fifty could expect to live another twenty-three years. So, despite the improvements in health care in the intervening time, life expectancy had increased by just three years.'[3] One can well believe that in the 1840s, those tough enough to survive all the diseases which removed half the population before they were fifty might well survive a good while longer. But the chance of reaching the age of fifty was far lower than today. The expectation of life at birth was then of the order of 42 years, some 30 years less than the expectation today. So, not only has life expectancy increased, but also the quality of that life has been improved, because so many disabilities have been removed. The only real interest in the quotation is in the question of how long we wish to seek to live, and in the intriguing scientific question of what biological process it is that actually limits the duration of our lives. Our attitude to the ideal length of life perceptibly changes as our age approaches (or passes) the accepted critical zone.

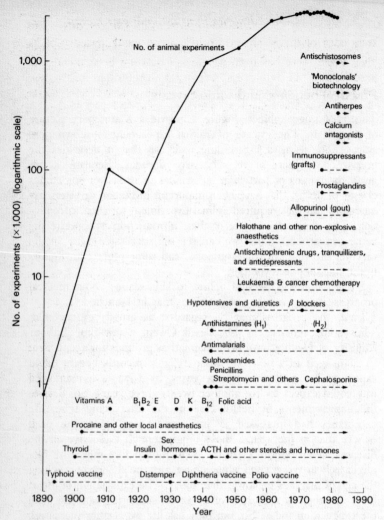

**Fig. 1.** Animal experiment and medical advance

The filled circle on the dotted line below each word or phrase describing each advance is set opposite the approximate date of introduction of that advance. Broken lines indicate continuous development.

The growth of animal experiment since 1890 is also shown, using a logarithmic (proportional) scale of the total number of experiments performed in the UK. The numbers are plotted at wide intervals of time until 1970, after which yearly figures are given.

Second, a good deal of the criticism of medicine suggests that, while there has been progress, it has not been therapeutic discoveries, but rather improvements in nutrition or hygiene or preventive medicine or changes in the characteristics of the diseases themselves that have led to the longer life and diminished morbidity of Western life today. This criticism was made particularly in relation to tuberculosis, and will be taken up when its story is discussed later in the chapter. It must be stressed, however, that much of the knowledge that made possible effective hygiene and good nutrition itself depended on animal experiment. Thus, to make sure that water is free from cholera, salmonella (the cause of gastro-enteritis), or legionnaires' disease organisms, or that milk is free from tuberculosis, one must already know that such organisms cause disease, and how to identify them. All this knowledge stems back to the work of Robert Koch and others in the last century, working out how to determine whether a particular micro-organism was harmful or harmless, in particular by showing in animals that a specific disease was *caused* by a specific organism. To this day, organisms are found in the body that are merely chance contaminants. The accounts of tetanus and poliomyelitis, below, illustrate the approach, and the controversy over the causation of AIDS emphasizes its importance. Equally, in nutrition, it required careful animal experiment first to *prove* that there was something in fresh vegetables and citrus fruits or in fat or in rice husks that was essential for human health. The work of James Lind and others, pointing to the importance of fresh fruit in preventing scurvy, was vitally important; but it was not until experiment made it possible to identify vitamin C as such that clear nutritional advice and implementation were achieved. Similar experiment has been needed to identify other vitamins so that they could be synthesized for therapeutic use and the amount in different foods determined. Nevertheless, if we trace the figures for growth of population and personal income over the centuries, we can readily recognize how mortality increases as pressure of population exceeds the availability of food, housing, and clean water and how it improves as personal resources increase.

It is no purpose of this book to argue that all advances in human health have stemmed from medical discovery based on animal experiment. The argument is much more important. It is that human and animal health depend on many things, including scientific discovery; that scientific work includes many branches, including animal experiment; and that it is a tragic error to try to set these various activities against each other.

When we seek to assess the part played by animal experiment, then, we must allow for these other factors. Equally, we need to be critical before agreeing that a real benefit has been achieved. It is not enough, for instance, to accept dogmatic medical statements that this or that drug is valuable. The most vivid illustration of this that I know is the opinion of J. A. Paris, who went on to become President of the Royal College of Physicians from 1844 to 1856, that calomel, soon to be condemned and now abandoned, was outstanding as 'a preparation more extensively and more usefully employed than almost any other article in the whole range of the materia medica'.[4] The tracing of the rise and fall of gold therapy for tuberculosis between 1924 and 1944 by Hart provides another example.[5] Today we recognize explicitly the way the course of a disease may fluctuate. In the past, the chance taking of a remedy just before a remission seemed conclusive evidence of curative effect; we are more cautious now. We are equally critical of patients' reports, now that we know of the 'placebo response'. In this, with almost any disease assessed chiefly by patients' testimony (common cold, 'rheumatism', indigestion), apparent relief in 30–50 per cent of cases can be obtained, for a while, with distilled water or sugar. We can understand the placebo response better, too, as a result of experimental work on what has been termed 'sensory decision' theory. This recognizes that a report of pain or suffering in response to some stimulus can be dissected into two components: (1) the actual ability to distinguish between the strengths of a series of varying stimuli—this is termed 'discriminability', and is thought to depend primarily on sensory nerve function; (2) the liability to classify a stimulus as having been painful—this is 'response bias', and is sensitive to expectation, suggestion, motivation, and the like. All four of the possible patterns are found: high and low discriminability with high and low response bias. Perhaps it needs to be pointed out that we can all be 'placebo reactors'; there is some physiological sense in the body having mechanisms whereby the significance of a sensation can be modulated by circumstance, even if it may complicate clinical trial.

Our first procedure, then, may well be to look for mortality or other general statistics, taking careful account of previous trends and other factors. We shall see some abrupt changes in these statistics, but the evidence will not rest on that alone. There will also be the knowledge—for instance, about the sulphonamides and antibiotics—that the drug is known to be experimentally effective *in vitro*, with a rational mechanism of action; that potency *in vivo* (estimated by blood levels) and *in vitro* correspond; and that individual case histories and analyses exist that convincingly illustrate the benefit.

One great difficulty remains: this is that we are largely restricted to statistics on *mortality*. The measurement of *morbidity* (suffering, illness, disability, handicap) in quantitative terms is extremely difficult, and reliable, large-scale data are lacking. So here we will have to leave out of account (and it is a big loss) all the relief given to, for example, sufferers from hay fever, insomniacs, the arthritic, the epileptic, the itching, the disfigured, and much mental disease. The reader, however, may be well able to fill in with the experience of his or her own family and friends.

Finally, it is worth recapitulating the tests to be satisfied before claiming a medical advance (here restricted to drug therapy) in which animal experiment has been an essential element:

(1) The essential link with animal experiment must be demonstrable. This requires scientifically informed historical study. Normally there will also be essential clinical and other contributions. (See Chapter 6.)

(2) There should be evidence of a real improvement in mortality, best at the level of population statistics, but often from a controlled clinical trial.

(3) Other influences (housing, wealth, population, etc.) must be allowed for.

(4) Mere opinion, however eminent or sincere, is not to be relied on.

(5) Allowance must be made for the possibility of spontaneous remission or recovery.

(6) Allowance must be made for the placebo response and its influence on the liability to report improvement or worsening. Parallel control, as in well-conducted clinical trials, deals with tests 3–6.

(7) As far as possible, some quantitative estimate of the benefit

must be obtained. There is nothing like quantitative measurement for sharpening the wits, calling of bluffs, and setting things in proportion.

(8) It should be verified that the drug concerned is known experimentally to be effective, with a rational mechanism of action within the framework of existing knowledge.

(9) It should be verified that the drug was given in amounts that would give blood levels likely to be effective.

(10) It should be verified that the benefit shows itself in individual case histories and tests, as well as in group statistics.

With surgical and other treatments, of course, other criteria would be appropriate. Criticisms of clinical medicine are usually accompanied by claims for other 'alternative' or 'complementary' therapies. It is no part of the purpose of this book to discuss these; but the criteria set out above are of the type that these other systems should also be required to meet.

## Benefits to man

We may begin with some examples from bacterial disease.

First, **puerperal sepsis**. Figure 2 shows the statistics for maternal deaths and puerperal sepsis (childbed fever due to streptococcal infection of the genital tract), with a dramatic fall beginning about 1935 when sulphonamides were introduced, to be followed in 1940 by penicillin. As a result, maternal mortality due to puerperal sepsis fell from around 200 per 100,000 births up until 1935 to 70—one-third of that figure—by 1940 and only about 5 in the 1960s.[6]

Second, **lobar pneumonia**. Pneumonia has a variety of causes, including viruses, whose treatment is still inadequate. But Figure 3, giving deaths specifically from lobar pneumonia (a very characteristic consolidation of the lung caused by the pneumococcus) in middle-aged men, shows the abrupt fall from around 60 per 100,000 to about 6 in 1970.[7]

Third, **rheumatic fever**. The full results of bacterial infection do not always show themselves during the acute stage of the illness. An important cause of heart disease in the younger age-groups used to be infection by a particular strain of streptococcus: for instance, by an attack of tonsillitis or scarlet fever. Some

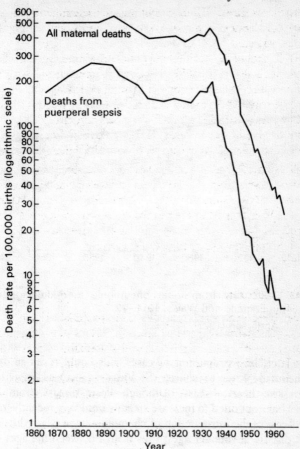

**Fig. 2.** Maternal death rate per 100,000 births in England and Wales, 1860–1964: all deaths and deaths from puerperal sepsis

Ten-year averages are given between 1861 and 1890, five-year averages between 1891 and 1930, and annual rates between 1931 and 1964.

Derived from the Registrar General's Decennial Supplement, England and Wales, 1931, Part 3, and the Registrar General's Statistical Review of England and Wales, various years. From the Office of Health Economics (1966).

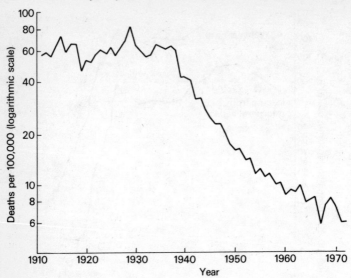

**Fig. 3.** Death rate from lobar pneumonia in middle-aged men (45–64) in England and Wales, 1911–72

From T. Anderson (1977).

years later, in a proportion of cases, this could result in attacks of rheumatic fever or chorea (St Vitus's dance) associated with progressive heart disease. Although heart disease from other causes has continued to increase steadily until very recently, one of the effects of the introduction of chemotherapy was a reduction of heart disease at younger ages; thus deaths between the ages of 15 and 24 fall from around 200 per million in 1935 to a tenth of that thirty years later.[8] Also there was a corresponding reduction in the occurrence of chronic disability in those who did not die. But in developing countries it is still a major problem and responsible for a third of heart disease.[9]

Next, let us consider some examples of the use of vaccines and antitoxins. First **diphtheria**, for which, first, an antitoxin was discovered to treat the disease, and then a vaccine to prevent it. Figure 4 shows the data, and also raises a valuable point about how one uses statistics. The result of these two advances was the

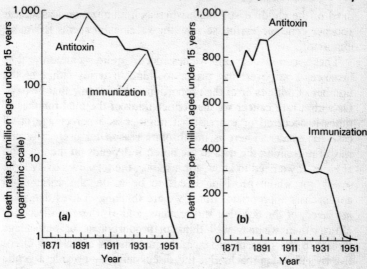

**Fig. 4.** Childhood mortality from diphtheria in England and Wales, 1871–1951: on a logarithmic scale (*left*) and a linear scale (*right*) From Altman (1980).

decline of childhood deaths from diphtheria from levels of around 800 per million towards the end of the last century to almost zero today.

Now we can think of such a reduction in death rate in two ways, which is worth a digression. A natural way is to count the lives saved—in other words, to use an *arithmetical* scale. But that says nothing about how many lives were at risk or how many lives were *not* saved. So another way is to express the saving of life as a percentage of those at risk—that is, to use a *proportionate*, or *logarithmic*, scale. The data in Figure 4 have been shown in both ways, arithmetical on the right, logarithmic on the left. From the right-hand graph we can see that antitoxin saved around 600 deaths per million children each year, while immunization 'only' saved the remaining 300 deaths per million; thus the antitoxin seems the more important advance. But if we turn to the left-hand graph, we see that even after the use of the antitoxin, about one-third of the children still died; whereas immunization was almost 100 per cent successful; so now the vaccination seems the more important.

These contrasting approaches occur quite commonly. For instance, a taxpayer may prefer to think in terms either of the number of pounds or of the proportion of his income that he pays. Or with a rare cancer we may either think of the total number of times it occurred or express that number as a percentage of all cases of cancer. There is no absolute reason for preferring one way of presenting the data to the other; it depends on the question at issue. If we need to know, for instance, the total lives likely to be saved, for whom provision needs to be made, the arithmetical approach is appropriate. But if we are thinking of lives that were *not* saved, of the task that still remains, and of where to direct our efforts, then we may well think in proportionate or percentage terms. In the present instance, the treatments are clearly effective by either approach. But the discussion may provide a gentle reminder to consider *how* such data are expressed.

**Tuberculosis** provides a fascinating case. The epidemiologist T. McKeown was very impressed by the fall in mortality during the nineteenth century, before medical measures such as vaccination and chemotherapy had had their full impact, and he argued that the benefit from medical measures had been overestimated.[10] He assessed the effect of streptomycin as follows. First, he took the death rates from tuberculosis since records began (1848). Then, because the population was always changing, he applied these rates to a standard population (that of 1901). He then computed the deaths that had occurred from 1848 up to 1971. For the years 1948–71, after streptomycin was introduced, he assumed that in its absence the death rate would have followed the trend of the previous 25 years.

The result, for the period 1948–71, was an estimate of 51 per cent lives saved by the treatment; for the whole 123-year period from 1848, however, the saving had only been 3.2 per cent (139,836 deaths prevented out of a calculated 4,377,265). It is this calculation, and similar treatments of other diseases and of gross mortality, that have been much stressed by critics of medical research and the pharmaceutical industry. No doubt, if one carried the research back to 1066, one could show that almost any change that took place was of trivial significance.

The reasons for the fall in mortality over the whole period are fascinating and much debated. A lengthy discussion by a leading textbook of bacteriology[11] refers to many factors: better housing and sanitary conditions, improved social habits, less drunkenness, better diet, higher standard of living, earlier diagnosis, segregation of advanced cases, sanatorium treatment, and the development of group immunity. As they remark, it is extremely difficult to assess the importance of these factors.

But despite the century-long fall, in 1950 there were still over 50,000 new cases of tuberculosis annually in the UK. For these, the effect of chemotherapy and then of Bacille Calmette-Guerin (BCG) vaccine was not a mere debating point (see Figure 5).

Figure 6 may also give some flavour of the change. Streptomycin had been discovered in the USA, and used rather indiscriminately. As a result, bacterial resistance rapidly appeared, and there were serious doubts as to how best to use it. The Medical Research Council (MRC) trials instituted in 1948 are classic for their methodology, revealing unequivocally the benefit from streptomycin and how resistance could best be minimized, by combination of streptomycin with p-amino-salicylic acid (PAS) or isonazid.[12] The results, partly shown in Figure 6, were conclusive; they had their own control group, and did not depend on assumptions about trends.

Tuberculosis remains a serious problem in many countries, and reappears in the UK from time to time. But we are no longer at the mercy of random forces; when it appears, the vaccine and drugs now available have made it both preventable and treatable.

**Tetanus** has been known for centuries, and a vivid description was given by Hippocrates. The excruciatingly painful cramps, during which, as in strychnine poisoning, the patient is fully conscious, gave it a particular horror, even though it was not very common. The bacterium responsible, with its spores, occurs in the faeces of domestic animals—for example, in horse manure—so that although it is commonest in rural and poorer countries, it remains a risk in gardening or farming communities. In the last century, the organism was identified by finding that injection of earth into animals could produce a disease like tetanus, and that from the site of the injection a characteristic organism could be obtained, which, if cultured, could produce the disease. It was then found that the bacterium produced an extremely potent toxin;

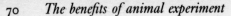

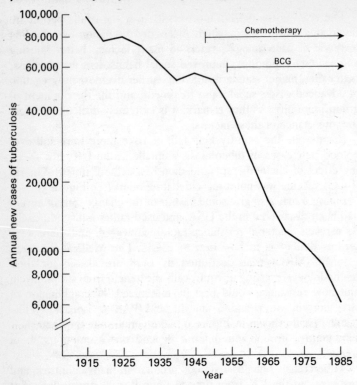

**Fig. 5.** Annual numbers of new cases of tuberculosis in England and Wales, 1913–87 (five-year averages)

From Hart (1988: 19–20).

that an antitoxin is formed in the animal body, which could be used as an immediate antidote; and that an inactivated toxin could be very effective in producing immunity. Animal experiment was necessary for all these discoveries. The wartime introduction of the antitoxin (crude though it was by modern standards) in November 1914 led to a dramatic fall in cases and in mortality.[13] More generally, deaths in the UK, which used to run around 150 a year, have now declined to 10–20. Vaccination against tetanus toxin is now a routine part of childhood protection. Equally significant has been the advance in treatment of tetanus,

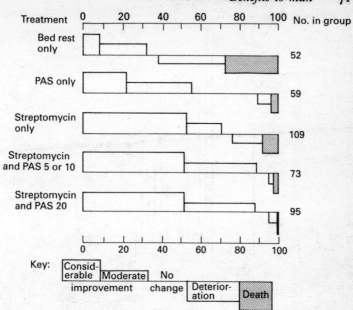

**Fig. 6.** Result of controlled clinical trial of streptomycin and PAS on pulmonary tuberculosis

X-ray assessment at 6 months; combined results of three trials. From (1962).

based on animal experiment, in those cases that still occur. As with strychnine, the painful convulsions can exhaust and finally suspend respiration. But the spasms can be abolished by the notorious curare, through its ability to paralyse muscles. This removes the pain and the exhaustion, but, of course, also affects the respiratory muscles. Artificial respiration, at which modern anaesthesia is adept, must therefore be provided. Other drugs are available to control overactivity of the sympathetic nervous system.[14] In a severe infection caught late, death still occurs; but the terror of the disease is now removable. To quote from one trial: '10 p.m., patient restless, crying out, tearful and frightened. After injection, quiet, rational, grateful.'[15]

Since the experimental discovery that the toxin acts very specifically in the nervous system by preventing transmitter release

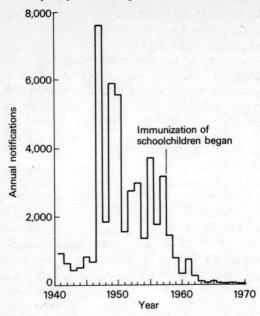

**Fig. 7.** Annual notifications of acute paralytic polio in England and Wales, 1940–70

From Paton (1979) and by courtesy of Professor H. Barcroft.

at inhibitory synapses, it has become an important tool in analysing mechanisms of central transmission. Another remarkable discovery, whose significance remains to be unravelled, is that the toxin reaches the central nervous system from a wound, not via the circulation, but by travelling along the motor nerves of the muscles back up into the spinal cord.

**Poliomyelitis** is another important case.[16] Figure 7 is interesting in illustrating another point in the handling of data. One might, for instance, simply have taken, as baseline, data from some date (say 1946 or 1955) during one of the epidemics, so as to make the improvement as dramatic as possible. But a wider look is more instructive: for, if we move the baseline back to a period between epidemics, we see that the introduction of polio vaccines has virtually removed not only the epidemics but also the background incidence from which the epidemics grew.

Poliomyelitis is a significant case in another way, in that it is very clearly not one of the diseases of which the incidence is reduced simply by raising the standard of living.[17] In countries with a low level of sanitation, the disease is mainly of the non-paralytic form, and is prevalent in the young. As sanitation improves, the age at which infection occurs comes later, and paralysis is commoner. The disease is commoner in the upper than in the lower social classes of a population. Incidence of the paralytic form has seemed to be greater the less crowded a household. Generally speaking, the more primitive the sanitary conditions, the earlier and more widespread is the development of antibodies. The original belief that the virus was spread principally by droplet infection was too narrow a view; it is now known to occur in faeces and to be found in sewage. It is now unequivocally established that vaccination by inactivated or attenuated virus, with its production of protective antibodies, is highly effective in practice in preventing the disease. It therefore seems clear that in conditions of low hygiene, much of the population was in fact receiving an uncontrolled dose of the virus in early life, giving for the most part lasting protection at the expense of a minor feverish illness. A higher standard of living, with its accompanying more elaborate hygiene, can be seen, ironically or even tragically, as removing this protection.

Poliomyelitis is also a striking example of the role played by animal experiment. Although the disease had been clearly identified by 1840, there was little understanding of it until 1909, when it was shown that an extract of spinal cord from a fatal case could transmit the disease when injected into a monkey. It then at last became possible to test for the presence of the infection, to show that it was carried by an agent that could pass through bacterial filters, and to begin to study its transmission and general properties. It also became possible to start finding out how to grow it in tissue culture and to characterize it as a virus, although it was not until 1949 that this succeeded. There soon followed a simple way of detecting in a culture the presence of the virus (e.g. by its damage to certain cells); methods of culturing the virus from the body, from faeces, and from sewage; methods of production,

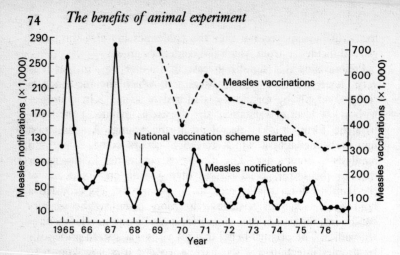

**Fig. 8.** Notifications of measles in quarterly periods October 1964 to December 1976 and measles vaccinations 1968–76 in England and Wales

From Medical Research Council (1977).

attenuation, and inactivation; and methods of identifying particular strains using specific antibodies. It is a classic example of one type of medical progress, from clinical medicine via animal experiment to effective prevention.

**Measles** is another childhood infection preventable by vaccination. Figure 8, from a Medical Research Council Report, gives a vivid picture of how the introduction of vaccination in 1968, by building up a good proportion of protected children within a year, almost abolished the biennial epidemics, as well as greatly reducing the background incidence.[18] The reduction in measles also meant a fall in the childhood bronchopneumonia to which it readily gave rise.

**Whooping cough** (pertussis) vaccine has been the subject of considerable debate. This is no surprise to the pharmacologist, who has used the toxicity of the killed organism for decades as an experimental tool.

The damage that a micro-organism does can arise in either or both of two ways. It may produce and release a poison into the blood—an 'exotoxin'; diphtheria is an example, and that is why the antitoxin was so successful. Alternatively, the body of the micro-organism, after it has multiplied and then died, may be poisonous (an 'endotoxin'); the whooping cough organism is one of these. Now the body of the micro-organism also contains the materials that produce immunity to the disease, and vaccines were made originally by simply injecting preparations made from dead organisms. If they are themselves poisonous, it obviously becomes necessary to find ways of separating the endotoxin from the constituent that produces immunity; and this offered some difficulty with the pertussis organism. In the process, it was necessary, of course, to have some measure of the effect of the endotoxin, so that one could be sure that it had been got rid of as far as possible; and it was from such work that its properties came to light. The whole story provides a reminder that a vaccine against some infectious organism is not, so to speak, automatically available; one must separate the toxin properly from the antigen; and until one can do this, vaccination itself may incur the danger of disease.

For pertussis the difficulty of doing this threatened to create a dilemma calling for the wisdom of Solomon; for it seemed that whooping cough vaccination carried a significant risk of causing damage to the brain. But it is also true that whooping cough itself carries comparably serious risks to the brain. In a six-month study during the 1974–5 epidemic, 10 children died and 1 child in 10 was ill enough to be admitted to hospital. In the 1977–9 epidemic, of those admitted, 12 per cent had pneumonia and 5 per cent convulsions.[19] But despite this, there was a real dilemma; for an unvaccinated child might not get whooping cough, and if it did, it would not be any *particular* person's fault. But vaccine brain damage would have been a result of the parents' and doctor's decision, and they would find it hard to feel free from blame.

In the event, however, it appears that the risk of nervous damage has been much overestimated, chiefly perhaps because it has been too readily assumed that *any* signs of some effect on the brain after vaccination (usually a multiple vaccination) can be due *only* to the whooping cough component and that there was no intervening disease of any sort. After a seven-year study of adverse effects of vaccination by the Public Health Laboratory Service Epidemiological Research Laboratory, involving 400,000 injections of vaccine, there was no evidence of brain damage due to it.[20] As the report says, 'If this syndrome exists, instances would probably have been discovered.' The acid comment at that time by

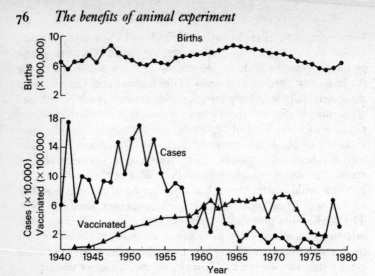

**Fig. 9.** Whooping cough incidence in England and Wales in relation to vaccination, 1940–80

The annual births are included, since this influences the number of cases. From Anderson and May (1982).

'Minerva', in the *British Medical Journal*, seems justified: 'So what has the national press, radio and television had to say about this reassuring important study? Not a word; or if anything has appeared Minerva missed it.'[21]

Figure 9, from another paper, gives an interesting picture of the incidence of cases in relation to the records of vaccination: since the number of cases must depend on the number of children born, the annual births are also shown, from which the proportion vaccinated can be estimated.[22] A fall in whooping cough is seen from 1950 onwards, as the proportion of children vaccinated steadily climbed. Then, after near freedom from the disease for 10 years, from 1976, there was a resurgence, after a media campaign on the dangers of vaccination had led to a drop in vaccination rate to as low as 31 per cent. The result was more than 30 unnecessary child deaths in the following years. One can also say, after reviewing the evidence, that in principle whooping cough epidemics could be eliminated; Figure 10 illustrates this for Fiji.

The one great example of the elimination of a disease is, of

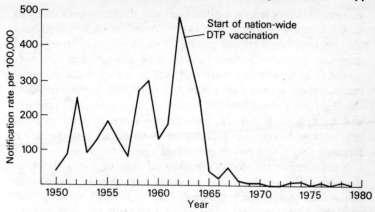

**Fig. 10.** Whooping cough notification rates per 100,000 population in Fiji, 1950–80

Vaccination acceptance rate in Fiji between 1972 and 1980 was about 85 per cent, calculated by subtracting infant deaths (5,782) from live births (162,430) and dividing the result into the number of completed courses of vaccine given to pre-school children (132,817). Vaccination was with a triple vaccine for diphtheria, tetanus, and pertussis (DTP). From Pollard (1983).

course, **smallpox**, and it needs no illustration. The last case was reported in Somalia, in 1977. But it is worth recalling that, for instance, death rates in India as a whole used to range up to 2,000 per million in a year, with far higher local incidences;[23] and that in those who did not die, the scarring for life with pockmarks could be a devastating disfigurement. It might be asked how far the use of animals contributed to this success, when one remembers that Edward Jenner's famous experiments on vaccination, recorded in 1798 in his *Inquiry into the Causes and Effects of variolae vaccinae*, were all on human subjects. For a long time, the principal use was as a standard source of cowpox lymph, raised on the abdominal skin of calves. But the abolition in the 1970s had to await the ability to identify viruses and for modern immunological knowledge, for both of which animal experiment was necessary in a variety of ways.

A lesser, but interesting, case is that of **dental caries**. There is an organism particularly associated with caries, named *Streptococcus* *mutans*. Monkeys fed on a high-sugar diet develop caries much as we do. It has been found, first, that a crude vaccine prepared from the organism could reduce this caries incidence considerably, and then that a purified antigen has the same ability.[24] One can only speculate as to whether there will ever be a case for widespread immunization against caries. But the work has thrown considerable light on a condition that causes much trouble and expense; and it emphasizes both the importance of oral hygiene and that general health and the capacity to generate an immunological response may play a role in protection against caries.

A more important area is the group of **chronic cardiovascular diseases** (Figure 11). One must expect chronic diseases to respond to therapy only gradually. But it is worth noticing that with the advent around 1950 of drug therapy for hypertensive disease—a treatment that was clumsy to begin with, but has got better and better over the years—the death rate from hypertensive disease has stopped rising, and is now steadily falling, together perhaps with the death rate from strokes, to which hypertension may lead.[25] In America there are now signs that death from coronary heart disease is also beginning to fall.

Finally, let us look at **childhood leukaemia**, and in particular the acute lymphocytic form. (In the older or middle-aged person it is commonly the chronic myeloid type.) Here I would like to illustrate how medical advance is often progressive—not so much a sudden 'breakthrough', but a succession of small advances that finally results in a major dividend.

Figure 12 records the first steps.[26] With no treatment, about 50 per cent of children under 15 with acute leukaemia were dead in 4 months, and all died within little more than a year. Then, around 1946–8, rather savage treatment with nitrogen mustard added perhaps a month to this time. Then some early anti-cancer drugs, which interfere with one of a cancer cell's natural growth factors (folic acid), were introduced (this type of drug is called an 'anti-metabolite'). These folic acid antagonists, together with steroids, lengthened survival to about 9 months; and with some newer anti-metabolites a 50 per cent survival of 12 months could be achieved by 1953. A recent study (Figure 13) shows the next step. Here the data are expressed in a different way, using not death rate but the presence or absence of leukaemia cells in the blood as an indicator (a 'haematological remission' is when they disappear after treat-

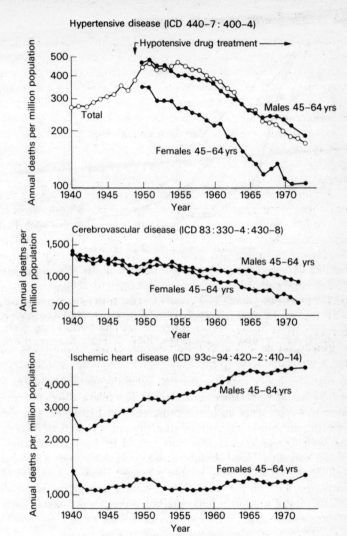

**Fig. 11.** Deaths from hypertensive, cerebrovascular, and ischemic disease in England and Wales, 1940–73

The International Classification of Diseases (ICD) was revised in 1949–50, 1957–8, and 1967–8, and the ICD categories used are shown. Death rates, adjusted for change in the age and sex distribution in the population, have been extracted for males and females of 45–64 years of age.

Data are from the Registrar General's Statistical Reviews of England and Wales. From Paton *et al.* (1978).

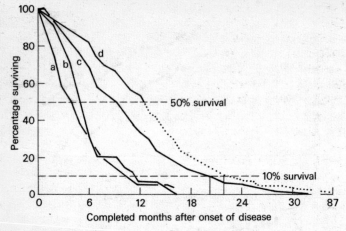

**Fig. 12.** Survival rates in the USA of patients under 15 years of age with acute leukaemia

(a) 218 untreated cases; (b) 34 cases treated with nitrogen mustard, 1946–8; (c) 160 cases treated with folic acid antagonists and/or steroids, 1948–52; (d) 184 cases treated with folic acid antagonists, steroids, mercaptopurine, azaserine, 1952–5. From Burchenal and Krakoff (1956).

ment). The upper figure shows that such a remission would now last for 12 months; since death would not follow till some time later, this represents further improvement on Figure 12. But the lower figure shows the result of an important new step, irradiating the skull to attack leukaemia cells lodged in the brain; because of the so-called blood–brain barrier, it is hard to reach them with drugs, yet they could initiate a relapse. With this additional procedure, over 60 per cent of patients were in remission for 100 months; and it is thought today that more than 50 per cent of such patients can be cured if they are under the age of 10.

## Cancer

This leads us to one of the largest fields of animal experiment, that of cancer research, which deserves more detailed discussion. Although it may seem pedantic, it is worth explaining some of the words used. 'Neoplasm', as its English translation 'new growth'

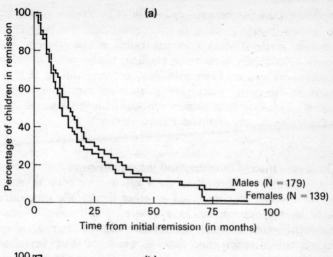

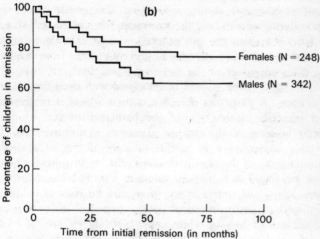

**Fig. 13.** Haematological remission in boys and girls in the USA under 16 years of age with leukaemia

In the first study (a) various methods of chemotherapy were used without any special treatment of the brain. The second study (b) was designed specifically to test the further protection achieved by irradiating the brain. N is the number of children treated. From Sather *et al*. (1981).

implies, means an abnormal growth of cells. Such a growth may be termed 'benign', like a wart, or 'malignant' if it is locally or generally invasive, when it is also called a 'cancer'. The word 'tumour', originally meaning a swelling, is also often used as a synonym for cancer. Even a benign neoplasm can be fatal if it grows in the wrong place; so the statistics may refer to 'neo-plasms', including both benign and malignant growths, although the latter are far the commoner cause of death.

## Cancer research

The research seeks to understand the conditions of tumour initia-tion, growth, and spread; to find selective methods of attack; and to develop effective drugs or other therapeutic procedures. Some of the experiments have aroused considerable criticism, particularly those in which tumours are allowed to grow, causing obvious animal misery, and those in which the drugs under test against an experimental tumour are toxic. Cancer also causes very great suffering in man, and the treatment, too, can be very severe. The latter is because the only generally successful strategy thus far has been to find drugs effective in preventing cells from dividing. This takes advantage of the fact that accelerated cell division is found in some tumours, and is associated with their growth and invasiveness. A particular objective is then to seek a preferential effect on cells of the type of the particular cancer, which is difficult because of their many likenesses to normal cells. An inevitable consequence is liability to some degree of a similar action on some of the rapidly dividing cells in our normal tissues. These are found in hair, skin, stomach, intestine, blood-forming organs, testis, and ovary; hence there may be risks of baldness, skin disorders, gastro-intestinal upset, anaemia, liability to infec-tion because of loss of white cells, bleeding by loss of blood platelets, and sterility. All this would be too heavy a cost of treatment for a minor disease, but is justifiable for a cancer. Cancers vary in their responsiveness to treatment, so a difficult clinical judgement may have to be made between the chance of palliation or cure and the deterioration of quality of life during treatment. Some especially successful areas are mentioned later. It also raises the question, for research, of how far animal models of disease may be pushed. Some may say not so far as this. Others may feel that, since domestic pets and other animals die of cancer

**Table 1.** Annual deaths from cancer per million population in England and Wales, 1871–1904

| 1871–5 | 1876–80 | 1881–5 | 1886–90 | 1891–5 | 1896–1900 | 1901–4 |
|--------|---------|--------|---------|--------|-----------|--------|
| 445 | 493 | 547 | 631 | 711 | 800 | 861 |

too, and since humans currently experience anything that animals are exposed to, it is legitimate, provided (as we shall review later) the science is sound and the suffering reduced to what is unavoidable.

Before considering the progress achieved, we must face the claim that is sometimes made that cancer mortality, far from having been reduced, has substantially increased despite all the research. This has prompted some to call for such research to be stopped. The point is an old one. Table 1 shows the figures given in an article in 1911 for crude annual death rates from cancer per million people in England and Wales.[27] But the shrewd comment was made that it was not clear how far this doubling in overall death rate represented a real increase in cancer incidence and mortality, how far it was due to more accurate diagnosis and improved recording, or how far it was simply due to the increased length of life, so that a larger proportion of people reached the age at which cancer is most frequent. Today such crude figures for death rate are no longer used, but only figures adjusted to a form not materially affected by the age distribution of the population as a whole. The usual form is that of *age-specific death rates* for particular types of cancer (i.e. rates in some narrow range of age—useful in detailed study of a particular cancer, but giving an overwhelming mass of data if applied to a population over a wide range of cancers); and *age-standardized rates* (where the rates are recalculated, knowing the age distribution of the population from which they derive, to give the rates that would be found in some standard population; for instance, Doll and Peto have used the US 1970 census figures as the best available standard.[28] Once the changes in age distribution due to changes in birth rate and longevity are pointed out, the fallacies in using crude death rates are obvious, but their magnitude will be less familiar. A few examples may be quoted from Doll and Peto. They describe some changes in age distribution in the USA between 1953–7 and 1973–7: for white males, the population of age 0–4 declined by 18 per cent; between 20 and 24 rose by 72 per cent (the result of a surge in the birth rate); between 40 and 44 was almost unchanged; between 60 and 64 rose by 26 per cent; and between 80 and 84 rose by 78 per cent. There were similar changes for white females, with a rise in those over 85 by 225 per cent. Non-whites show general rises at all ages, but also the same increase in the older groups. To such causes of misunderstanding must be added allowance for errors or changes in diagnosis, registration, and sampling, each a matter of considerable study. The most reliable statistics date from 1950; with care, data may be used back to 1933; before that the uncertainties are too great. All this produces a greater caution in dogmatic statement, the more expert the investigator.

Some conclusions, despite these difficulties, are possible.[29] First, on the general trend of age-standardized cancer mortality: if one sets aside cancers related to smoking (in the respiratory and upper alimentary tract), the trends in mortality over the most recent 10-year period for some 30 groups of cancers in both sexes are mostly downwards, by amounts generally of the order of 1–3 per cent a year. The chief exceptions are pancreatic cancer in women and melanoma in both sexes.

Second, to know if this is due to improvement in cure rate, it is necessary to have information about cancer incidence. Here the statistical data are inevitably subject to even greater difficulties of interpretation. But a cautious interpretation of the evidence does point to a general small improvement in 'relative survival rates' (a relative survival of 100 per cent would indicate a risk of death the same as that of the standard US population as a whole matched for age and sex): thus the relative survival rate has risen (from 60 per cent in 1950–4 to 68 per cent in 1970–3 for cancer of the female breast, and there are similar changes with all the common cancers except stomach, pancreas, and lung. There is greater success in three areas:

(a) **Hodgkin's disease**, where survival over the same periods rose from 28 per cent and 34 per cent to 66 per cent and 69 per cent for males and females respectively.

(b) **Leukaemias** in the young, where three-year relative

**Table 2.** The growing contribution of lung cancer to total deaths from neoplasms: deaths per 100 million US population under 65 years of age

|  | 1935 | 1955 | 1978 |
|---|---|---|---|
| **Males** | | | |
| All neoplasms except respiratory | 58,709 | 57,176 | 52,538 |
| Respiratory cancer | 5,812 | 19,495 | 33,816 |
| **Females** | | | |
| All neoplasms except respiratory | 90,108 | 71,893 | 58,618 |
| Respiratory cancer | 2,102 | 2,820 | 12,064 |

*Source*: Doll and Peto 1981, Table D1, adapted

survival had risen from 5 per cent in 1950–9 to 24 per cent in 1967–73 for those under 35 years of age. Figures 12 and 13 show reason for believing that population statistics for childhood leukaemia today may show a still greater hope of survival.

(c) The less common **cancer of the testis** and **choriocarcinoma** have also shown major improvement in survival rate.

**Lung cancer** requires special discussion, as the outstanding exception to the general drift downwards in cancer death rates. The figures in Table 2 make the point.

It is now generally accepted both that the association between cigarette smoking and lung cancer is real and that it is causal. In what way, if any, has animal experiment contributed to the recognition of this cause, to the understanding of the mechanism involved, or to the prospect of reducing the mortality? The recognition of a causal association must certainly be attributed primarily to skilful and tenacious research by epidemiologists, studying the incidence of disease in social groups or whole populations. But it seems that the acceptance of such a conclusion by the general public also requires some understanding of the mechanism involved. This can be illustrated by the history of research into puerperal sepsis. The Hungarian investigator Semmelweis had proved by around 1850 that childbed fever was produced by the transmission of infective material by accoucheurs into the maternal genital tract from the post-mortem room or from other patients. Yet no one believed him, and he died a disappointed man. Just 20 years later Lister's antiseptic method was winning international recognition. The main reason why Lister's conclusions were accepted, whereas Semmelweis's were rejected was the effective birth of bacteriology in the intervening period. No longer was it necessary to postulate an unknown, hypothetical infective material; instead, there were microscopically visible organisms. The mechanism linking cigarette smoking and lung cancer is still by no means clear. But our understanding has progressed considerably. Examples are the explicit demonstration that cigarette tar painted on a mouse's skin can cause cancer, the chemical identification of highly potent chemical carcinogens in tars (the benzpyrene series), the recognition of the complexity of carcinogenesis and of the processes of 'initiation' and 'promotion', the discoveries of how these seemingly inert compounds are chemically transformed in the body into much more reactive substances, and the identification of a chemical reaction between these substances and the constituents of the cell's genetic material. These are only a few of the developments, directly or indirectly making the pattern of the mechanism more intelligible.

There has been one further step. The animal work has indicated that it is not primarily the nicotine nor the gases nor the carbon monoxide in cigarette smoke that is carcinogenic, but something in the tar, and there has been a great deal of research into its composition and properties. Over the last two decades, the level of tar in cigarettes has been steadily lowered. Much research has also gone into exploring tobacco substitutes which could lower the risks of smoking by lowering the yield of the various harmful substances. The epidemiological results, just beginning to emerge, are most encouraging. The effect is most obvious, as one would expect, in younger members of the population, for whom the 'lower tar era' represents a larger part of their lifetime experience of smoking. Table 3 shows some of the results.

Sir Richard Doll comments on this result that it

detracts in no way from the importance of education and other measures to discourage smoking altogether, which can eventually have a much greater effect on total mortality from a variety of cancers and other disease; but it is unrealistic to think that a habit that has become so commonplace and involves so many financial interests can be eliminated

**Table 3.** Annual deaths per million from lung cancer in England and Wales, 1940–80

| Age-group | Men born *c.*1910 observed 1940–60 | Men born 1930–50 observed 1980 | Percentage reduction |
|---|---|---|---|
| 30–4 | 40 | 13 | 64 |
| 35–9 | 98 | 45 | 54 |
| 40–4 | 293 | 134 | 47 |
| 45–9 | 597 | 378 | 37 |

*Source*: Doll 1983, Table IV, adapted

overnight. It is evident, however, that people are willing to switch to low-tar cigarettes and if, as now appears, such a switch leads to a major reduction in the risk of disease, we need to do everything we can to bring it about.[30]

The recognition of the role of smoking must be above all credited to the epidemiologists; but for their arguments to be accepted and for the low-tar cigarette to be introduced and accepted, there have been many other essential contributions, including that of animal experiment.

## Cancer due to occupational and environmental agents

There is now a formidable list of agents established as able to cause cancer in man. Among those arising in everyday life, as well as smoking, are aflatoxin (produced by mouldy food in the tropics), alcohol, chewing betel, tobacco, and lime, certain parasites (for example, bladder cancer from schistosomes), ultraviolet light, and hepatitis B virus infection. Industrial work may involve exposure to a carcinogen: for example, to asbestos, benzene, cadmium, some factors in furniture or leather manufacturing, nickel, beta-naphthylamine and other aromatic amines, polycyclic hydrocarbons, and vinyl chloride. Some substances used in medical treatment or in self-medication may also be carcinogenic, such as oestrogens, phenacetin, steroids, immunosuppressants in organ grafting, and alkylating agents. Their initial recognition arose in various ways from case history, epidemiology, and animal work. But with them all, the establishment and acceptance of the cause, the understanding we have, and the means of control rest on no single fact or discipline, but on co-ordinated work in almost all branches of biology and medicine.

The demands for control, together with the number of substances, are largely responsible for a great deal of industrial testing of new compounds. Mere enumeration has begun to give the impression that it is substances of this type that cause the bulk of human cancer. But it is an important conclusion, in Doll and Peto's remarkable and convincing book, that all these industrial compounds together probably account for no more than 10 per cent of human cancer deaths, while tobacco and dietary factors are responsible for two-thirds, and viruses for a substantial proportion more. This conclusion calls into question the extent and nature of the testing of new substances, so extensively required of industry at present. Certain simple policies must clearly be rejected, however. To stop all testing, including animal tests, and to rely solely on epidemiology would result in thousands of avoidable deaths, and is clearly unacceptable. Public policy over the nitrosamines already illustrates the rejection of such an approach. Here are a group of chemicals that were found by animal experiment to be among the most potent carcinogens known in animals. As a result of this finding, and in advance of any epidemiological proof of a hazard in man, determined efforts are now made to reduce nitrosamine levels in food and elsewhere in the environment as far as possible.

Nor can one rely solely on non-animal methods of testing before human exposure is allowed. There are a variety of such tests: tests on enzymes, on the liability to mutate in bacterial culture (the Ames test), and on cultured mammalian cells. Alternatively, quick animal tests involving little suffering may allow recognition of an interaction with DNA or of a genetic change or of an effect recognized microscopically on reproductive cells. Such tests have the great advantage of speed and of avoiding long-term animal exposure. Yet their predictive power is quite clearly not strong enough; and indeed, it is unreasonable to expect such tests over a few days or weeks to predict what will happen over the whole of a human life-span. At the same time, long-term animal tests are also limited in predictive capacity, so that one cannot rely on them alone either.

There is evidently no simple answer. There is a lot to be said for Doll and Peto's suggestions for estimating which substances are most likely to matter: they propose using laboratory tests for what they call 'priority setting', rather than for formal 'risk assessment', and of then combining *in vitro* and animal tests with epidemiological estimates of frequency and intensity of exposure. Testing for carcinogenicity is discussed in Chapter 10. A rather hopeful trend in these directions, together with new genetic approaches, is in fact developing.

## Advances in surgery

The account so far given has focused chiefly on medical conditions and on treatment with drugs and vaccines. This is both because of my own experience and because mortality statistics of the kind so far cited at the population level are less readily available for surgical procedures. But we must nevertheless review some uses of animal experiment in surgery.

Once anaesthesia and the control of infection had opened the gate to advances in surgery, animal experiment began to find the best material to use for one of the surgeon's primary tasks—rejoining tissues that he has had to cut through. Experimental wounds were made under anaesthesia, and then repaired with the test material. A wide range of **suture materials** has been tried: hemp, silk, catgut, various metals, tendon, horsehair, even strips of skin or blood vessels. The necessary qualities were gradually discovered. The suture material needs to be strong enough to hold, for instance, the abdominal wall together, even after exposure to the enzymes in the tissues. It needs to maintain its strength even in a moving organ such as the intestine or when exposed to such fluids as gastric juice. It should be able, ultimately, to be reabsorbed, yet should maintain its strength until the repair mechanisms of the body have allowed scar tissue of sufficient strength to form. It must be sterilizable, non-irritant, and non-antigenic. The actual loads to which a suture may be subjected had to be measured. A special problem was that of how to suture, in a child, an organ that was still growing. One of the most delicate recent achievements is that of the microsurgery of blood vessels, with the development of materials and techniques that allow operation on blood vessels no more than 1 mm. in diameter. All this called for decades of experiment ranging from Lister's original studies in the last century up to the present day.[31]

A second major area lies within **orthopaedic surgery**. Before modern techniques could be developed, physiological experiment was necessary to discover—for instance by using a dye (madder) that is taken up only by growing bone—that the long bones grow, not evenly along their length, but from their ends; that regeneration of bone occurs from the tough surface layer over the bone called the periosteum; and that the body can 'remodel' bone by combining bone removal with bone deposition.[32] Ways were found to graft bone, to ensure the sterility of the grafted materials, and to preserve such material in a hospital 'bone bank' for future use. Armed with such knowledge, it became possible to correct skeletal deformity and, for instance, to hold back the growth of a normal limb in a child so that, in adult life, it should be the same length as a limb whose growth has been impaired.

A great field of advance has been in surgery of the hip joint. A common accident in the elderly is to break the neck of the femoral bone where it is set into the hip socket. A solution is to reset the fracture and then hammer a metal pin through the neck to support it. But some metals, when surrounded by tissue fluids, are liable to become electric batteries, and to set up electric currents which can dissolve bone and produce other changes that destroy the firmness of the support. So metals had to be found that were 'electrolytically inert' and that were compatible with the tissues. These metals could also then be used as splints to broken bones, so that some jockeys and racing motorcyclists are now liable to be walking ironmongery shops!

Surgery of the hip has gone still further in patients with osteo-arthritis, where the whole hip-joint is replaced with an artificial socket embedded in the hip-bone, into which fits an artificial femoral head embedded in the femur. Again, compatibility with the tissues had to be secured. For this, modern plastic has proved of great use; and the success of the whole research is shown by the waiting-lists for 'artificial hips'. There is, in fact, a growing field of implant materials, as new uses multiply. These are not only for joints or heart bypasses, but also for drug-delivery devices, blood filters, the artificial pancreas, and the like.[33]

Another important area is **cardiovascular surgery**.[34] As Comroe and Dripps have shown in detail, this has been built on centuries of experimental and clinical work on both the normal function of heart and circulation and their disorders. Today an artery blocked by arteriosclerosis can be replaced or bypassed, either by a vein or by artificial materials, backed by knowledge of what happens to these not just immediately after the operation, but over the following weeks and months. The **cardiac pacemaker** is another great advance. It depends both on the knowledge of how to make an implant compatible with the tissues and on a deep knowledge of the electric currents that flow when the heart beats and of how the beat itself is initiated. **Cardiac surgery** only became possible (except in rare cases) with the development of two techniques: the heart-lung machine, which allowed the circulation to vital organs to be maintained while the heart was stopped, and deliberate hypothermia, where the body temperature is cooled so low and the need for oxygen by the brain and other organs is thereby so reduced that an absence of blood-flow can be tolerated for the duration of an operation. Together with great surgical imagination and skill, this allowed operations on so-called blue babies, in whom congenital abnormality of the heart threatens the life of an otherwise normal child. Along with this went the design and testing of artificial heart valves to replace those damaged by disease and the working out of procedures for stopping a beating heart before operation and of restarting it afterwards. Yet another promising new approach is to use a back muscle, after it has been appropriately mobilized and electrically 'trained', to boost a failing left ventricle. The possibility comes from physiological work on the adaptation of muscles to their tasks. Originally, this related to the way that muscles used for rapid movement were largely white, whereas those used to maintain posture were red, with a variety of other deeper differences. Now a muscle previously used in the body for intermittent body movement can be trained to become efficient at rhythmic contraction.[35]

The heart-lung machine has a sister in the **artificial kidney**. Failure of the kidneys can arise in many ways: for instance, from disease, from poisons or wrongly used drugs, or from traumatic shock such as accompanies gross crushing injury. The idea of 'renal dialysis'—that is, taking blood out of an artery, passing it over a thin membrane with a suitable salt solution the other side to let waste products diffuse away, and then returning it to the body—is about 70 years old. It only needs a glance at a microscopic section of a kidney to understand why it has taken so long to understand the physiological processes well enough to be able to make an artificial substitute. Eventual success has, in fact, created a dilemma. For while such artificial kidneys are satisfactory when kidney failure is only temporary, the cost of lifetime provision for many thousands of patients is very large indeed.

This has proved a great stimulus to **renal transplants**.[36] This, too, was envisaged by surgeons several decades ago; but it needed an understanding of immunology, of the physiological 'laws' which determine whether a grafted tissue or organ will be accepted or rejected by the body, before it became practicable. The problem was somewhat analogous to that of blood transfusion, for which a donor's blood group must be matched to that of the recipient; but the expression of individual characteristics in a whole organ is very much more complicated. The matching of what are called 'histocompatibility antigens' has greatly advanced. This, together with experimental development of ways of storing kidneys from dead people and of drugs which can hold back immunological rejection (the so-called immunosuppressants) have opened the way to renal transplant as a general technique. If successful—and success rates are steadily rising—the operation constitutes in effect a cure, and the patient is freed both from the constraints and the costs of regular renal dialysis. As so often in medicine, the success has come from a succession of small steps, in improved immunology, better surgery, and better drugs, depending alike on clinical, animal, biochemical, and chemical experimental work. A similar account could be given of **corneal grafting**, first explored with rabbits' eyes. The same gradual improvement is visible with grafts of heart and lung, liver, pancreas, bone marrow, cartilage, and nervous tissue.

We could summarize what animal experiment, with other work, has offered to surgery by reviewing what would happen, say, to a soldier with gross injuries to the face from bomb blast. Blood loss could be compensated for by blood transfusion, and infection could be combatted by antibiotics. Surgical operation would be under a safe anaesthetic combined with analgesics if necessary. The suture material used would leave the minimum scarring. Skin or bone graft could help in reconstructing the face. If he could not swallow, artificial feeding of a carefully worked out nutrient 'cocktail' would be infused into a vein. If shock had produced renal failure, some form of dialysis could be used. The dressings applied would prevent infection, be non-irritant, prevent fluid loss, and foster healing. None of these procedures would have been possible without animal experiment.

## The change in appearances of human disease

Our review of some of the historical benefits for which animal work was essential has been chiefly statistical. But to someone who was trained in medicine 50 years ago, an equally cogent and more vivid argument is the simple abolition of some everyday manifestations of disease. One no longer sees infants with ears streaming pus, schoolboys with facial impetigo, beards growing from heavily infected skin, faces pocked by smallpox or eroded by lupus, or heads and necks scarred from boils or suppurating glands. Drugs and a better diet have transformed haggard patients with peptic ulcers. The languid, characteristically brown-skinned case of Addison's disease of the adrenals; the pale, listless patients of chronic iron deficiency or pernicious anaemia; and the cretin or, conversely, the young woman with 'pop eyes' and overactive emotional behaviour—due respectively to thyroid deficiency or excess—are all being treated. The soggy hulk of a patient in the oedematous stage of chronic kidney disease is relieved by diuretics. As a result of polio vaccine and control of tuberculosis, we see few crippled children; as one walks behind a group of youngsters today, varied as ever in shape and size, the marvel is how straight their limbs and backs are. The chronic arthritics with their sticks are being replaced by septuagenarians swinging along on their plastic hips. The patients now are rare that once one saw dying from an infected mastoid, struggling for breath in the last stages of heart failure, or dying from appendicitis, leukaemia, pneumonia, or bacterial endocarditis. For those without such memories, the pictures of Hieronymus Bosch or the cartoons of Rowlandson or Hogarth will give some inkling. Equally vivid, for those with some medical knowledge, are Professor Paul Beeson's accounts of how therapy has changed in the last half century and of how these changes were brought about.[37]

It is worth recalling these appearances, not in the interests of complacency, but as a reminder that at the time it seemed that one could often see no chance of doing more than ease the patient's path and hope for nature's curative powers. Sir William Osler, whose scepticism of most of the remedies of his day helped to create an atmosphere of 'therapeutic nihilism', wrote in 1909:

To accept a great group of maladies, against which we have never had and can scarcely hope to have curative measure, makes some men as sensitive as though we were ourselves responsible for their existence. Those very cases are 'rocks of offence' to many good fellows whose moral decline dates from the rash promise to cure. We work by wit, not witchcraft, and while these patients have our tenderest care, and we must do what is best for the relief of our sufferings, we should not bring the art of medicine into disrepute by quack-like promises to heal, or wire-drawn attempts to cure in what old Burton calls 'continuate and inexorable maladies'.[38]

Each generation confronts problems that seem to it insoluble. Yet the recollection of diseases once regarded as 'continuate and inexorable maladies' that have now disappeared teaches us to look in a different way at diseases which today seem equally inexorable. If, as Thomas Browne wrote about the memory of those who have died, 'oblivion blindly scattereth her poppy',[39] so too, Nature seems blindly to scatter her seed-corn of future discovery. As we look at the intractable problems of today—chronic neurological disease, schizophrenia, senile dementia, most forms of cancer, congenital deformity,[40] deafness, blindness, arthritis, auto-immune diseases, and many forms of tropical and veterinary disease, as well as emerging new infections[41]—it seems right to say *not* that they are insoluble, but that it is our task to seek the knowledge which may provide the seeds of solution either in this or in a later generation.

## Benefits to animals

One of the really remarkable features of the antivivisectionist literature is the lack of mention of the benefit to animals that medical research has brought. Many of the drugs and procedures that have been of importance for man are of equal effectiveness in animals. The veterinary surgeon has the same antibiotics and antiseptics; the same hormones, tranquillizers, local anaesthetics, and general anaesthetics; the same surgical procedures and principles of resuscitation or other life support; and many of the same nutritional supplements. Indeed, the techniques of anaesthesia are so alike that human and veterinary anaesthetists co-operate in professional training and operational research.[42] If one compares the medical and veterinary pharmacopoeias,[43] over half of those in the veterinary pharmacopoeia come from human medicine; the remainder focus on the specifically animal infections and parasites. The veterinary pharmacopoeia should be compulsory reading for those doubting the animal benefit from animal research! Table 4 shows some of the drugs shared by human and veterinary pharmacopoeias, and Table 5 some of the vaccines developed for animal diseases. James Herriot's stories about the life of a veterinary surgeon provide many illustrations, some of them all the more vivid because he was practising while the therapeutic revolution was coming in.[44] This extends even to relatively rare conditions: in dogs, bladder stone due to a congenital cystinuria has been controlled by the human remedy penicillamine; the identification of a type of haemophilia in man with that in dogs has advanced knowledge for both, and there are many other similar cases of similar inherited disorders.[45]

But there are two main areas specific to animals, where the work was done specifically to deal with animal disease: with major epidemic diseases of various sorts and with infection by worms. The attempt to produce vaccines against animal disease began almost as early as that for human disease, with Pasteur's work both on anthrax and on rabies. Fifty years ago Sir Leonard Rogers was able to calculate that by then over 100 million animals had been saved by inoculation against anthrax and rinderpest (cattle plague) and a similar number by swine-erysipelas inoculation in Germany since the treatments were discovered.[46] Dog distemper, which used to kill animals by the hundreds of thousands, has also been nearly abolished. Other epidemic diseases due to organisms related to typhoid, dysentery, plague, cholera, and gas gangrene have been controlled. Virus diseases that can be checked by vaccines include Marek's disease and a feline enteritis. Both have proved important in understanding the relation of viruses to carcinogenicity; indeed, the feline infection is in some respects a model for AIDS.[47] The *British Pharmacopoeia (Veterinary)* now contains over 30 vaccines, and new developments continue. A recent one is the discovery of a drug treatment for an African disease, 'East Coast fever', due to *Theileria parva*, carried by ticks; this has been killing half a million cattle a year by a febrile disease affecting the lungs and ending with the animal drowning in its own

**Table 4.** Some drugs developed for human use employed in veterinary medicine

*Antibiotics and anti-infectives*
Penicillins
Cotrimoxazole
Chloramphenicol
Erythromycin
Metronidazole
Neomycin
Sulphonamides
Tetracyclines
Streptomycin

*Gastro-intestinal*
Bismuth salts
Electrolyte solutions
Hyoscine
Paraffin
Sulphonamides

*Circulation*
Digitalis
Glyceryl trinitrate
Isoxupren
Frusemide
Thiazides

*Epilepsy, sedation, anti-emetic*
Phenytoin
Primidone
Acetylpromazine
Fentanyl
Acetylpromazine
Metoclopramide

*Pain and fever*
Butorphanol
Pethidine
Phenylbutazone
Meclofenamic acid
Paracetamol + codeine +/−
    butobarbitone

*Nutrition*
Essential fatty acids
Vitamins A, B complex, C, D, E,
    K, folic acid, biotin, selenium
Iron dextran

*Anti-inflammatory*
Betamethasone
Dexamethasone
Prednisone
Triamcinolone

*Worms and parasites*
Thiabendazole
Pyrantel
Praziquantel
Piperazine
Ivermectin

*Lung and allergy*
Sodium cromoglycate
Theophylline
Tripelennamine

*General and local anaesthesia*
Halothane
Thiopentone
Ketamine
Pentobarbitone
Propofol
Lignocaine
Procaine
Mepivacaine

*Eye and skin*
Antibiotic eye lotions
Antiseptics
Griseofulvin
Permethrin

*Hormones*
Chorionic gonadotrophin
Oestradiol
Progesterone
Oxytocin

**Table 5.** Vaccines for animal disease

|  | Viruses | Bacteria |
|---|---|---|
| Dogs | Distemper<br>Hepatitis<br>Para-influenza<br>Rabies | Leptospirosis (Weil's disease)<br>Bronchitis<br>Gammaglobulin |
| Cats | Feline enteritis<br>  (panleucopenia)<br>Herpes<br>Rhinotracheitis<br>Parvovirus<br>Rabies |  |
| Piglets and sows | Rotavirus | Enterotoxins (*E. Coli* strains)<br>Bordetella (bronchitis)<br>Erysipelas<br>Pig cholera |
| Sheep |  | Enterotoxins<br>Clostridia (blackleg, braxy,<br>  footrot, pulpy kidney)<br>Salmonella |
| Calves and Cattle |  | Clostridia<br>Salmonella<br>Parasitic bronchitis |
| Horses | Equine herpes<br>Equine influenza<br>Other viruses | Tetanus antitoxin<br>Tetanus toxoid<br>Pasteurella |
| Birds | Paramyxovirus<br>  (pigeons)<br>Herpes<br>Bursal disease<br>Encephalomyelitis<br>Newcastle disease | Pasteurella |
| Fish |  | Furunculosis (Aeromonas) |

froth. A more sophisticated advance is the combination of animal work and molecular biology to produce a new vaccine for foot-and-mouth disease.

The other main area of disease is responsible as much for chronic illness as for mortality: namely, infection with nematodes (roundworm), cestodes (tapeworm), and trematodes (the flukes). The roundworm can both damage the alimentary tract and get into the lungs; the tapeworm causes wasting; and the flukes, with their extraordinary life cycle, can produce very varied damage. For almost all of them today some sort of treatment is possible; and the greatest practical problem is how to make a treatment that is equivalent to the deworming of a domestic pet available on a far larger scale.

The reader, after reviewing such evidence, even if he acknowledges the transformation in the life of livestock and domestic pets, may nevertheless feel that animals' interests are still not taken properly into account and that the benefit is essentially for an unworthy cause—namely, to make it more possible for man to use animals more readily for his own purposes—as food, wool, skins, for sport, or as pets. Yet whatever happens to the animal in the end, it must be worth while to make its life more comfortable; what we *use* animals for is a separate issue.

A similar argument could be advanced against the sulphonamides and penicillin. At one time, epidemics of cerebrospinal fever (meningitis) used to occur in young soldiers clustered together in barracks, and could restrict the concentration of troops at some point at the front. This was one way in which the therapeutic revolution aided the world's armies, as typhoid vaccination had done earlier. Yet who would turn the clock back because it helped the soldier as well as the citizen, just because the soldier's job is to fight? It is the same argument that would have led to the withholding of anaesthesia for childbirth, because the suffering was considered at one time to be essential to maternal moral welfare. It could also be argued, along the same lines, that in order to prevent promiscuity, chemotherapy should not be used to treat venereal disease. In all these cases there is a suffering human being or animal: we may wish they would go away, but in the real world, where one sees the ravages of cattle plague, the suffering of meningitis or of prolonged, difficult childbirth, or the end results of venereal disease, who can do other than seek to remove them?

## Benefits in tropical medicine

Anyone dealing with the benefit from animal experiment or from medical research as a whole for tropical medicine must do so circumspectly. A routine jibe is that all that medicine has done is to substitute death by starvation for death by disease. The short answer is that a relatively healthy population is more likely to find ways of improving its economy than a diseased one. A more serious comment would be that made by L. G. Goodwin in his two tables (Tables 6 and 7 below) listing respectively the six diseases earmarked for special attention by the World Health Organization and the proposed uses for the 61 new drugs developed in one particular year (1975)—none having anything to do with tropical medicine.[48] That must dispel any complacency, and the reasons are not hard to find. While the strategic knowledge required for a medical advance may come from university, institute, or industry, the actual drug development under today's conditions can come only from industry. It is difficult, unpredictable, extremely expensive, and takes between five and seven years; the product may be pirated by those who have done nothing to develop it; and the people who need the drugs are too poor to pay for them. So the field is hardly viable financially, and all research proposals must pay their way, whether to a nationalized or a

**Table 6.** Six diseases selected for special attention by the World Health Organization

| Disease | Number of people infected (millions) |
| --- | --- |
| Malaria | 200 |
| Schistosomiasis | 200 |
| Filariasis | 300 |
| Trypanosomiasis | 8 |
| Leishmaniasis | 2 |
| Leprosy | 11 |

**Table 7.** New drugs, by class, introduced in 1975 (in those countries chiefly concerned with drug development, especially USA and Western Europe)

| Therapeutic use | Number |
| --- | --- |
| Central nervous system | 15 |
| Anti-infective | 12 (none for tropical disease) |
| Cardiovascular | 7 |
| Anti-inflammatory | 6 |
| Cancer | 6 |
| Hormones | 4 |
| Respiration | 3 |
| Gastro-intestinal | 3 |
| Others (including diagnostic) | 7 |

private industry. The problem, however, is like that of the other 'orphan' drugs for diseases too rare to provide a sufficient market to support production, so that in effect a subsidy is required. Fortunately it is at last being taken seriously at the only level at which it can be dealt with—that of co-operative international governmental action.

Given that background, the success achieved in tropical medicine may seem all the more notable. Some of the successes are very familiar to the traveller: vaccines for smallpox, yellow fever, typhoid and para-typhoid, and cholera (rather poor initially, but now improving). Considerable help has been provided by establishing the nature of the vectors (insect or animal) which transmit a disease; and this can allow control, even if treatment is inadequate. Drug treatment for a long time relied on traditional remedies: quinine for malaria and ipecacuanha (emetine) for amoebic dysentery. But the Second World War provided a great stimulus, and new antimalarials, together with DDT to control the insect vector of typhus, sulphaguanidine for dysentery, and dapsone for leprosy, led the way. Further antimalarials followed, with drugs for filarial infections, for schistosomiasis, and for kala-azar, trypanosomiasis, and onchocerciasis. Their success is more impressive, however, if one looks at the past rather than at what remains to be done. Drug resistance is emerging; some of the agents are distinctly toxic; and a pervading difficulty, almost as great as that of discovering a new drug, is that of discovering how to get medical care delivered to those who need it. Indeed, the latter applies, too, to all those diseases that the industrialized world shares with the developing world; for instance, there are the usual bacterial infections. Prevalence of rheumatic fever remains high in India, Africa, the Philippines, Central America, Indonesia, and the Middle East.[49] Measles attack rates of nearly 50 per cent have been recorded in young children in The Gambia, although it was shown to be reduced to 5 per cent after vaccination.[50] Poliomyelitis still paralyses seven in every 1,000 Ghanaian children.[51] Much more could be done simply by implementing existing knowledge more thoroughly.

Nevertheless, the outlook for tropical medicine has, in principle, been revolutionized. The *possibility* of control of almost any of the diseases concerned is now quite clear, and even where resistance develops, our understanding of the cellular mechanisms involved is already pointing to ways of counter-attack. In all this, animal experiment, along with clinical medicine, epidemiology, zoology, chemistry and biochemistry, and engineering, have all played their part.

We should note, too, the extent to which man has used himself as an experimental subject in such work. Humans have volunteered to be infected with yellow fever, poliomyelitis, malaria, infectious hepatitis, as well as with less hazardous yet hardly more appetizing roundworms and threadworms. A remarkable modern instance is the work by Ralph Lainson and his colleagues on leishmaniasis in South America.[52] The task they set themselves was to get convincing evidence about the vectors—in other words, how the disease was carried. This they did first by acting as human bait for sandflies (standing at night stripped to the waist and then collecting the various species). They then found which species could carry the organism, and surveyed its natural occurrence. They learnt how to keep the insects and to make them carriers, and studied how the insects could acquire the organism from existing skin lesions. Finally they produced an infection with one of the infected sandfly species in a volunteer. Figure 14 shows a sketch of the organism's life cycle. The work was combined with similar work using hamsters and other rodents, in whom leishmaniasis occurs naturally. It constitutes an extraordinary

**Fig. 14.** The ecology of leishmaniasis of the skin in Brazil

Sandflies (*Lutzomyia* species) which have not yet laid eggs (NV) ascend tree trunks to the canopy where they feed at night (V1, V2) on primary 'reservoirs' such as sloth and anteater (1). Opossums (2) and a third sandfly (V3) may sometimes be involved. Engorged flies migrate down trees to lay eggs (oviposit: 0). During the day they move back to the canopy for the next blood meal, when infected flies transmit the infecting organisms (*L. braziliensis guyanensis*). If disturbed at ground level during the daytime or early night-time, some flies will feed on the nearest mammal, and terrestrial species (2t) may be infected. Man is a common 'victim' and host (3) because of his excessive disturbance of the environment. Broken lines represent transmission to subsidiary or 'victim' hosts. From Lainson (1982).

detective story in the maze of subtle adaptations between various species of organism, insect, and animal.

## Limited experiment and unlimited benefit

We must remember, too, one consequence of man's accumulative capacity. His successors tomorrow build on what he discovers today. It is true that ideas, theories, and interpretations evolve:

some of the science of today will seem as bizarre to our successors as the humours or the old phlogiston theory of combustion do to us. But that does not destroy the value of the knowledge won. While Einstein may have corrected Newton, yet for the vast mass of everyday practice, Newton's laws are sufficient. Our understanding of what happens when we vaccinate against a disease is sure to change. But the knowledge that enabled us to abolish smallpox does not have to be rediscovered, nor how to anaesthetize, nor how to kill organisms with antiseptics and chemotherapy, nor how to repair our damaged joints, relieve the asthmatic, or ease pain. If any suffering was entailed by making such discoveries, it is over; but the benefit to knowledge and to practice stretches far into the future for millions of humans and animals yet to be born. Every year adds to that tally of misery, suffering, and death prevented.

It is a powerful and a dangerous argument; for it might seem that it could justify almost any treatment of animals, or indeed of man. Three things should be pointed out, however. First, it brings out the well-known weakness in simple utilitarianism, which means that it is never in fact practised; it is always tempered, though in ways that are not unanimously defined, by other aspects of what is right and permissible. Like all moral decisions, it remains for individuals to take, in their own social milieu. Second, it brings out how immensely worth while medical discovery, humanely achieved, can be; it long outlasts most of the world's goods and, indeed, human and animal lives. That is a reason why medical research is worth pursuing and supporting. Third, it represents a major stimulus to finding ways of humane discovery. Some of the most encouraging developments today are miniaturization, computers, and new sensing devices. These steadily increase the scope for really sensitive, rapid, fully controlled, humane investigation.

## The only useful suffering?

If we look at the world as it is, we can be overwhelmed by the suffering it reveals. But that is too one-sided a picture. Man's personal cruelty to man goes along with great generosity and kindness. War shows ruthlessness and brutality, but also great courage. Nature may be red in tooth and claw, and, as has been said, 'The end of every wild animal's life is a tragedy'; but there is also much evident beauty and enjoyment. Nevertheless, it is still true, as Lord Justice Moulton said, that 'The greater part of pain were better not to be.'[53] So much of it seems useless, even when one has allowed for need of warning of bodily damage or for the necessary price that freedom of action in a dangerous world calls for. It is all the more strange, therefore, that the suffering which carries the hope of reducing future suffering is so bitterly attacked. What *use* is there in the tens of thousands of cases of gratuitous cruelty recorded by the RSPCA every year? What *use* is there in the suffering as a cat plays with a bird or a dog worries a sheep or in any of nature's predatory activities? What *use* is the pain of a knee-capping in Northern Ireland or shattered limbs on a battle-field or of the mugged victim or of the traffic accident? Each points forward to yet more suffering, as revenge works on and while the puzzles of how to resolve conflict or to moderate human action remain unsolved. But if we turn instead to animal experi-ment and to the medical and veterinary practice built on it, it is a different picture, with some prospect, stretching into the future, of reducing human and animal misery. Without such experiment, without the deliberate search for knowledge, all we could look forward to is the long, random, incompetent trail of nature's own experiments on human and animal life.

## The 'test of deletion'

The test of deletion, introduced on p. 55, has been subject to two objections: (1) that logically one cannot predict the future from the past, and so cannot know the consequences if some discovery had not been made; (2) that one cannot know that the discovery would not soon have been made in a way not involving animals, especially when one considers all the new technologies available.

It is true that there are philosophical problems with 'induction', which hinge on questions of absolute certainty, rather than practical confidence. Thus there may be a philosophical uncertainty as to whether the sun will rise tomorrow. But everyday life is in fact conducted on the basis that it *is* possible to learn from experience and to act accordingly. Concern about relying simply on trends was, of course, the stimulus to the development of the controlled clinical trial. This reveals directly the progress of a disease in the absence of a remedy under trial. A result from such a trial was shown in the discussion of tuberculosis.

But controlled trials, which include a group of untreated patients, have obvious disadvantages at the population level, and then trends in population evidence have to be used. One could, of course, resort to detailed trend analysis; but usually it is clear enough whether some change is convincing or merely suggestive. If one considers the results for maternal mortality in Figure 2, for example, up to 1935 there is a general pattern of maternal mortality that had lasted for 70 years; few would fail to predict its continuance. It was in that year that the first chemotherapy, sulfonamides, was introduced. If one then looks at subsequent years, few would fail to recognize a real change, or be unwilling to use a comparison between the two periods to estimate 'lives saved'. One could, of course, search for other possible agents in 1935; but the time relationships are striking, and no alternative has been found.

The fact is that the 'test of deletion' is an entirely everyday process of mental comparison ('Goodness, that made a difference'). It has been introduced because of the curious neglect of the past in this field; it is taken too much for granted, and its lessons neglected.

The second objection suggests that such discoveries would be made in any case, in some other way, especially because of new technologies. The remark is vacuous, unless an explanation is offered: (a) of what that way is, or what the actual basis for any expectation of it is; (b) why it has not been found already; and (c) why people should have to continue waiting. None of this is done. It is true that there have been great advances in technology. But technology is not enough—indeed, it can be a hindrance when a scientist begins to tailor his enquiry to his technology, rather than his technology to his ideas. This may already be happening with some of the work on alternatives.

The reader is asked now to turn back to Figure 1 (p. 60) and to take a slip of paper which he can move along the time axis, starting from 1890, hiding what is to come. (The experiment is rather interesting.) The arguments against animal experiment have changed little in a hundred years, and if they are valid now, they would have been valid then. Let us suppose that you accepted them in the past, and that they would be enforced at your wish.

Where will you put your bar, or your restriction, to animal experiments? Which of the advances mentioned would you be willing to have dispensed with or to have delayed? What accompanying ignorance of the human or animal body, now removed, would you have wished to perpetuate? At each date you consider, the benefits and knowledge still to come were all hidden, just as future benefits are hidden today.[54] The only warranty for future work, then, lay in the historical record of what had already been achieved, the knowledge already gained, and the openings to new discovery that had been created. The position is the same today: the only difference lies in the richness of the past record.

## Summary

1. The 'benefit from knowledge' is assessed by imagining that no animal experiment had been done for 2,000 years, so that we lacked all knowledge so gained. The resulting ignorance both about our own bodies and those of animals would profoundly affect our outlook, on each other as well as on animals; it would remove much of the impulse to animal welfare, and, since knowledge about the inanimate world could have progressed largely unimpeded, it would produce a remarkable asymmetry in the pattern of our understanding—primitive as regards the animate, highly advanced for the inanimate. Deliberate failure to win new knowledge is the same as deliberate perpetuation of ignorance.

2. In assessing 'practical benefit' flowing from animal experiment, care is needed to allow for other causes of advance and for the lessons learnt about the placebo response and the fact that one cannot rely solely on doctors' or patients' approval of a remedy. A number of tests are reviewed. The lack of reliable morbidity data restricts us largely to records of changing mortality. Cogent illustrations of advances for which animal work was essential in a variety of diseases are cited, including infectious disease, cardiovascular disease, leukaemia and cancer, animal diseases, and tropical medicine.

3. One consequence of man's capacity to accumulate knowledge is that limited experiment gives rise to unlimited benefit: smallpox vaccine does not have to be discovered again. This introduces something in the nature of a multiplication factor in assessing benefit.

4. Of all the numerous forms of animal suffering that exist, only that incurred in animal experiment and the medical and veterinary practice based on it offers the prospect of *reducing* future human and animal misery; yet it is this that is most violently attacked.

5. A 'test of deletion' is outlined whereby the effect of abolishing or restricting animal experiment now is assessed by considering the effect of a similar abolition or restriction in the past.

# 6

# The pattern of discovery

The human mind is often so awkward and ill-regulated in the career of invention that it is at first diffident, and then despises itself. For it appears at first incredible that any such discovery should be made, and when it has been made, it appears incredible that it should so long have escaped men's research.

Francis Bacon, *Novum Organum*, Aphorism CX

The last chapter reviewed a small part of the benefits that have flowed from research to which animal experiment was an essential contributor. We may well admire and be grateful for it; yet we may still wonder what assurance there is for future success. Bacon's aphorism quoted above is profoundly true. What *has* been achieved soon seems easy, and is taken for granted; and only the teacher or historian gets insights into how difficult these achievements were. If we look to the unknown future, how are we to know, for instance, that *these* research workers provided with *these* resources will make responsible and successful use of them? It is, in fact, a familiar question: we ask, too, how confident are we that this doctor or this solicitor will give us good advice or that this architect will build us a good house or that this Member of Parliament will represent our interests? There, too, we look at their past records. But the question has a sharper edge when it concerns discovery, the bringing to light of what is, at present, unknown; and it is nearer to asking whether this musician or poet or painter is worth supporting. We have to face the fact that, in principle, we cannot say what will be discovered, or when.

Yet history is more reassuring than this. It is not simply that one can quote examples of the apparently trivial becoming important; such as the work on the pigments in the butterfly's wing (the pterins) becoming central to certain areas of biochemistry such as the synthesis of nucleic acids or the way that the exquisite sensitivity of the back muscles of a leech to nicotine provided the technical gateway to the theory of chemical transmission. It is rather that it is very hard to say of any piece of work that *no* practical benefit can be envisaged as resulting from it. Hearing me make this remark at a press conference once, a journalist challenged it, saying that he had done some graduate research on the biochemical uptake of an amino acid into the retinal pigment of the rabbit's eye and that it seemed to him to have been of little scientific and no practical use. Yet the answer was obvious. Either as a result of vitamin A deficiency or as a congenital defect, a condition known as 'night blindness' can occur, due to a defect in or absence of the retinal pigment. The metabolism of that amino acid, or some disorder of it, could easily become a valuable technique or clue in seeking to understand these conditions.

## The discovery–benefit lag time

Fortunately, we can turn from such anecdotal instances to an interesting and comprehensive study, illustrating how it is possible to do 'research on research'.[1] This was an investigation by two Americans, Julius Comroe, a cardiovascular physiologist, and Robert Dripps, an anaesthetist, both of considerable distinction, into the knowledge that led up to ten major advances in the medicine and surgery of the heart, circulation, and lungs. Their method was first to collect opinions from nearly 100 specialists about which had been the 10 most important advances in the field. Then, with the aid of over 160 consultants, they reviewed the body of knowledge that it had been necessary to acquire before the 10 clinical advances could reach their present state. In the process, they examined around 6,000 published articles, identifying about 3,400 specific scientific papers for study. Of these, 663 were regarded as key ones, underpinning the bodies of knowledge required, which fell into 137 separate lines of scientific investigation. It is worth giving some of their final conclusions:

(a) The public, including physicians and many scientists, still equate one important discovery with the name of a single man, e.g. polio vaccine = Salk. However in every instance that we studied, previous work by scores or hundreds of competent scientists was essential to provide the basic knowledge for the widely known clinical advance, usually attributed to one man.

(b) Of the 663 key articles essential for 10 major clinical advances, 41.6% reported research done by scientists whose goal at that time was *unrelated* to the later clinical advance; 41.6% sought that knowledge for the sake of knowledge. Such unrelated research was often unexpected, unpredictable, and usually greatly accelerated advance in many fields.

(c) Analysed in another way, of the 663 key articles, 61.5% described '*basic*' research, research that was performed to determine mechanisms by which living organisms (including man) function or by which drugs act; 20% reported descriptive clinical investigations without any experimental work on basic mechanisms; 16.5% were concerned with development of new apparatus, techniques, operations or procedures; 2% involved review and synthesis of earlier work.

(d) Of the key research, 67.4% was done in colleges and universities or their medical schools and associated hospitals.

(e) Although most key research was done in clinical or basic science departments of medical schools, important contributions came from all of the basic non-medical science disciplines in colleges and universities (biology, botany, chemistry, mathematics, physical chemistry, physics, plant physiology and zoology) and from agriculture, dentistry, engineering, photography, and veterinary medicine, as well as from a variety of industrial laboratories.

(f) Except in unusual instances (e.g. the clinical use of X-rays), some lag always occurred between an initial discovery and its effective clinical application. We analysed the length of 111 such lags: 8% amounted to 0.1–1 year; 18% were 1–10 years; 17% were 11–12 years; 39% were 21–50 years; only 18% required more than 50 years for application.

The study was a pioneering one, so far unique in its scale and care, and can of course be criticized. It is rather US-centred. The judgements were made essentially by scientists with little help from historians. Despite all precautions, arbitrariness could not be avoided. Different results could well be obtained in other fields. The calculation of percentages is probably misleading. Yet one need only review the sequence of scientific reports that they assembled to accept the general truth of their conclusions: that science is a co-operative enterprise; that 'basic' research is essential for practical advance, just as (although they were not investigating this) practical discoveries often have a 'basic' significance; and that the benefit of a discovery in one area of science often requires information from other areas of science before its potential can be developed or even envisaged.

It is worth quoting a few particular examples in order of increasing magnitude of lag time. The discovery of X-rays was reported by an academic physicist in December 1895. Already in the following year many investigators were using X-rays for medical diagnosis. By hindsight, though, one might say that their use spread *too* fast, as the later evidence of the damaging effect of over-exposure to X-rays became apparent.

A typical 'goal-oriented' discovery was halothane, developed as a general anaesthetic free from the explosion risk of ether, more potent than nitrous oxide, and safer than chloroform. Around 50 years of academic and practical work on men and animals preceded its chemical synthesis in 1951; it took another 5 years for its safety, its potency, and its special properties to be worked out in animals and for its production and formulation for clinical use. Today it is one of the most commonly used anaesthetics, in man and animals.

It used to be thought that it would be impossible to operate surgically within the abdomen, on the heart, or on the brain. The first repair of a wound in the heart, that of a rabbit, was made in 1882. It took 15 years to achieve the first success in man.

In 1922 a disease in cattle was identified: the main symptom was severe bleeding, and it was associated with the eating of spoiled hay that contained sweet clover (*Melilotus officinalis*). The chemical substance (dicoumarin) in the clover that was responsible was identified. It was found to antagonize the action of a vitamin—vitamin K—required by the body for making certain clotting factors. As has happened so often, a poison, once understood, offers the chance of being exploited as a therapeutic agent: in this case the possibility now existed of a drug to control blood-clotting in human disease, for which there was at the time no drug that could be taken by mouth. When tested, it proved reliably active by mouth in animals and man, and its duration of action and safety in other respects were mapped out. Then, 19 years after the starting-point, it received its trial as the first anticoagulant active by mouth, one of a group of substances now widely used to control blood-clotting and thrombosis.

The first sulphonamides (of which 'M & B 693', used for treating Winston Churchill's pneumonia during the War, is one of the most famous) were discovered in 1935. It was noticed by an investigator in 1937 that sulphanilamide produced a loss of sodium from the body. The trail then took many interesting turns.

The beginning—chemotherapy with careful clinical observation—was followed by basic work on an enzyme in the red blood cell; that enzyme was then found in stomach and kidney, and it was discovered that the sulphonamide could inhibit it. The mechanisms were studied; other factors controlling urinary secretion were recognized; and related chemical structures were tested. Finally, 29 years later, industry had developed an important series of safe, orally active diuretics, used for controlling excess water and salt in the body. Another important dimension was then added, as knowledge about the value of controlling raised blood-pressure developed, with the gradual recognition of the effectiveness of these diuretics for such control.

As an example of a very long lag, one may cite blood transfusion. It was first done in 1667, by an early Fellow of the Royal Society, Richard Lower, from one dog to another. The first *safe* transfusions date from the First World War, nearly 250 years later. What occupied the interval? It was soon found that transfusing animal blood into man was dangerous, and it was not until 1818 that James Blundell, having established that the transfusion must be from the same species, performed the first transfusion of human blood: it was in a dying patient, but produced only temporary improvement. His first successful transfusion, using blood given by his assistant, was in a case of haemorrhage after childbirth. But transfusion between humans also proved unpredictably dangerous, and further advances had to wait until a pioneer of immunology, Landsteiner, opened the way by discovering the blood groups in 1900. Then the discovery of reliable methods of blood storage and of preventing blood-clotting, the development of suitable equipment, and the organization (under the stimulus of war) of donor services and distribution allowed transfusion as we know it to develop.

Lastly, one may recall the work by Sir Henry Dale mentioned in Chapter 3. His study of the contaminants of ergot improved our understanding of shock, which was important in the First World War a few years later. It led to the establishment of the theory of chemical transmission in the 1930s, and was important for the development of antihistamines in the 1940s and 1970s and some major drugs for controlling blood-pressure in the 1960s—a whole family of discovery–benefit lag times.

As we look back over such cases in detail, each step clearly needed other advances, depending in their turn on yet others. We see, as the fundamental reason for such lags, that the different parts of science do not stand alone, and that not only is it impossible to foresee the pattern of growth in each area, but that the future interactions are also hidden.

Despite all this, the unpredictability must not be overstated. It is not true to say that the outcome of research cannot be foreseen at all, and therefore cannot ever be commissioned. A more accurate remark would be that the greater the originality of a piece of work, the less predictable it is. A great deal of research in fact yields results whose general character can be, if not predicted, at least named in advance among the possible outcomes. It is this, in fact, that provides the backbone of research, fundamental or applied: the gradual movement, guided by logic and imagination, from existing knowledge to the adjacent possibilities. The unexpected is the uncovenanted bonus for those ready to receive it. But the capacity to force one's way to a particular solution remains limited. My own best insight into this came from participating in personnel research into diving and submarine physiology during and after the War. When the War started, a diver or a crew seeking to escape from a sunken submarine was known to face certain hazards, particularly drowning, poisoning by oxygen or carbon dioxide, and decompression sickness; but the safe limits were unknown. Work on human volunteers established adequately for the first time the pattern of oxygen poisoning, and the amounts that could be breathed for a given time without symptoms or convulsions. It also discovered the amount of carbon dioxide required to produce unconsciousness. The design of breathing sets and the efficiency of soda-lime canisters for absorbing carbon dioxide were improved. Survival at sea was improved by appropriate study. Schedules were improved for slow ascent by a diver after particular tasks to avoid decompression sickness. But at the end of all the goal-oriented research, which resulted in a much safer diving practice, the great questions were still unanswered: how is it that oxygen is toxic or that carbon dioxide anaesthetizes or that gas dissolves in the body under pressure and then separates out so unpredictably after coming to the surface? As a result, practice remains empirical to this day, and the greater freedom that would result from being able to control toxicity or bubble formation still eludes us.

The great question of judgement, therefore, both for the larger issues of deployment of scientific resources and for the particular issue of the use of animal experiment, lies in the balance between the easily foreseen and the creation of openings for the unforeseen.

## The dangers of preoccupation with benefit

The argument has been presented so far as though interest in knowledge and interest in benefit were two fluids that could be mixed in any order and proportion. Yet there is a sense in which knowledge always has a priority. Both scientists and those who commission their work have particular hopes; and there is always the danger that the results seeming to fulfil those hopes will receive the greatest attention, or even be incorrectly selected as typical. The nature of scientific work ultimately provides a protection, for the scientist's search is for that which any other scientist can confirm, something independent of the observer and his own wishes. This is the fundamental importance of the confirmatory experiment, which would otherwise seem redundant. But formal, precise repetition of each others' work is fairly rare. The usual procedure is to test some possibility that would be the case if the original conception were true, and only if this fails, go back for a re-examination. So the possibility of a scientist moving ahead on insufficient evidence, believing he has found something 'useful', is only too real. The only protection is to forget about 'useful' and to ask simply, 'Is this the case or not?' Reliable benefit can only be built on secure knowledge.

The issue has come up in an interesting way in discussions of the ethics of clinical trial. The traditional procedure has been to set 'limits of statistical significance' to the results of such trial; that is, to say that knowing, for instance, how variable and unpredictable the course of some disease in a group of patients can be, one does not accept evidence of an improvement in rate of cure unless it is bigger than a certain magnitude.

For instance, suppose that with some disease it is known that, on average for a large number of patients, 50 per cent get better in 10 days. If you took a random sample of 10 such patients, the actual number getting better by that time would only sometimes be exactly 5, but would vary between 3 and 7 in most cases, according to the chance of whether favourable or unfavourable cases happened to be included. To be more exact, for trials of groups of 10, you would expect cure rates of between 34 and 66 per cent in two-thirds of the cases, and only 1 time in 20 would the cure rate be outside the range 18–82 per cent. So if you were trying a new drug on such a group, you would regard a cure rate of 66 per cent as no more than encouraging; but a 90 per cent cure would be quite convincing. (Equally, a cure rate of only 10 per cent would be convincing evidence that the drug was not merely useless but harmful.) Normally a difference is required greater than that which would be expected to occur 1 in 20 times by chance. (Those odds are not unreasonably strict: to toss 4 heads in a row—which many people will have done in an idle moment—represents a chance of 1 in 16). Yet it has been claimed that it is ethically wrong during a clinical trial, once *any* apparent advantage of one treatment over another (say 51 per cent cure against 50 per cent) has been shown, not to give to all patients thereafter the apparently better treatment. That particular case does not merit serious consideration; but as the apparent difference between the treatments increases, or as the odds of the treatments being equally effective move from near even to 1 in 3 and upwards, although still short of the 1 in 20 level, the question whether the trial should be abandoned becomes sharper and more awkward. So, too, those interested in animal welfare might ask whether animals could be saved by reducing the reliability of the knowledge to be gained; in fact, in a later section on toxicity testing such an argument will be seen to have something to offer.

The heart of the issue is probably quite simple: it is a question of deciding whether the experience gained is to be used as the basis of future action or not, and if so in what way. If firm knowledge is not an objective, if a physician is willing to use a treatment which is still quite likely to turn out to be the inferior one, if the scientist is willing to go ahead with a grossly unreliable index of toxicity, that is one thing. But if something is to be of use to others, if some security in one's knowledge for the future is required, there is no choice but to obtain that security.

## The whole animal as a superb detector

The reader may well accept the argument thus far: that discovery is hard to foresee, that science is really a family of sciences, that basic and applied research are both needed, that they interact in subtle and complex ways, and that one must not be impatient. You could still ask why it is that with so much knowledge won it is *still* necessary to do experiments on whole animals.

The easy, yet correct, answer is that living organisms are so complex that we are far from being able to solve our scientific problems on isolated parts. We shall return to this when we discuss alternatives (Chapter 8), but some explanation is desirable now. This complexity of living organisms arises in several ways. The full complexity can be expressed by the simple but rather intimidating statement that there are estimated to be of the order of 50,000 genes in the human genome, with possibilities of inter-action between them and between their products. But it is worth quoting some more straightforward examples.

First, the blood is a very complicated fluid. As well as carrying oxygen, it contains nutrients, required for the various tissues (absorbed from the intestine or already processed by one organ and being passed to another); hormones, whose amounts vary with the state of the body; controlling proteins of various kinds, binding or transporting various essential substances or controlling water flow to the tissues; and electrolytes (salts), controlling the responsiveness of the tissues, their water content, and their acidity. (The blood could, for a gardener, be compared to a particularly satisfactory circulating potting compost!)

Secondly, various tissues (especially liver, lung, kidney, and the lining of the small intestine) engage in a variety of biochemical transformations affecting the body at large: this may be the making available for other tissues of sugar, particular forms of fat, or special proteins; or it may be a protective mechanism, taking some substance which is potentially harmful and converting it to a less harmful form ready for elimination.

Thirdly, perhaps the most characteristic feature of a living organism is its power of adaptation. This is 'vitally' necessary in a literal sense; for the organism's continuing life depends on its capacity to maintain the special features of its normal existence— its body temperature, the correct supply of electrolytes and nutrients, and effective elimination of waste products. Any pertur-bation immediately leads to responses calculated to restore normality—the so-called homeostasis, or in the words of Claude Bernard, the French physiologist who first drew attention to it, 'the preservation of the internal environment'. These responses are extraordinarily varied: they may take place in seconds, like the blanching of the skin to minimize heat loss in the cold; or go on for weeks like the change in blood formation after bleeding or exposure to the lowered oxygen supply of high altitudes. They may involve almost any tissue, any hormone, or any part of the nervous system.

Finally, and especially in respect of the nervous system, there are profuse long-distance connections and interactions between the different parts of the body.

The consequence of all this is twofold. The first is the obvious point that isolation of any organ or any bit of tissue robs it of participation in this integrated physiology; and our success in making experiments on such isolated portions depends entirely on our capacity to replace artificially the tissue's normal needs and environment. The history of tissue culture records the progressive recognition, still incomplete, of all the conditions and all the nutrient substances required in the fluid bathing the cells. Even today, foetal calf serum is often required for the unidentified factors it contains favouring growth in tissue culture.

But a second consequence is that *it is the whole organism in particular that gives us our greatest chance of seeing the unexpected. It alone contains all the mechanisms, known and unknown, that we seek to understand*; it already has all the astonishingly sensitive recognition systems for the various known and unknown hormones and chemicals of the body; its adaptive responses represent an indicator system that something we may not have been aware of has changed. The more we move to work on isolated parts, the more we are restricted simply to those possibilities that we have thought of. It is because of this that the whole animal must be used for experimenting with a new drug or exploring the physiological action of some newly discovered body constituent or investigating the function of a newly identified gland or cluster of nerve cells in the brain.

## Individuality and variability

The other side of the coin of the complexity of living organisms is that each one is an individual, and that they therefore vary. With so many influences at work, as well as normal genetic variation, it is inevitable that responses to an experimental test will be different from one individual to another. This is what makes biological work fascinating; but it exacts a price. Where a chemist can measure the melting-point of a compound once and for all, the biologist will find, for instance, that each animal has a slightly different body temperature from every other, and that it varies with time of day, activity, and so on. To reach any general conclusion, he or she must therefore find ways of minimizing the variation and of allowing for it. It also makes confirmation by other investigators all the more important.

Biological and medical statisticians have in fact made major advances in statistical theory in tackling the problems created. The problems always arise in the same general form, involving three factors: the variability of the response being studied (which can sometimes be usefully broken down according to identifiable sources of such variability), the accuracy required, and the number of observations to be made. The variability of different biological characteristics is, in itself, very interesting. Sometimes the variation is small, like the level of sodium in the blood, which is regulated within 1–2 per cent. Other characteristics, like human height, vary, but only within 5–10 per cent—for instance, the British Association found that, for those born in 1883, two-thirds of adult males had heights between 64 and 71 inches. Yet other characteristics are very variable indeed, such as human hair length or male hairiness. The interest lies in the fact that if some characteristic is relatively constant, then there must be some biological mechanism regulating it rather tightly; so we have a direct indicator of the existence and effectiveness of the homeostasis mentioned earlier, and we can see, equally, that hair length is variable because there is little reason to control it (short of special Army discipline!).

The extent of the variability of a response and the precision required by the investigator are thus of key significance for animal experiment, essentially dictating the number of animals that must be used. Sir Richard Doll has remarked, 'There was a saying among my colleagues when I began the first year of preclinical studies to the effect that if it moves, it's biology; if it changes colour, it's chemistry; but if it doesn't work, it's physics. Some years later a friend added that if it sends you to sleep, it's statistics.'[2] Any teacher will confirm how much more than a grain of truth there is in this, but this book is no place for developing the statistical ideas that have been evolved by biomedical statisticians. It is, however, worth mentioning three points. First, a very great deal of effort, over 50 years or more, has gone into the task of discovering how to get the maximum of reliable information from a minimum of observations, both by finding ways of controlling variability or allowing for it and by using experimental designs with the best mathematical basis. Secondly, estimating the variability of a response, as well as being of some interest, is itself useful—it helps you to design later experiments better, and it can (for instance) warn a physician who is going to handle a new drug whether a small increase in dose is going to produce a big or only a small increase in effect. Thirdly, in the end there is no avoiding the fact that the precision of a conclusion and the number of observations made have to be traded off against each other: if you wish to be very precise and to have confidence in that precision, biological variation is such that a good many experiments must be done. The point is probably already familiar to the reader from experience with opinion polls.

## Some pessimists

It is not surprising that those attacking animal experiment should be critical of the quality of the work done. But it is curious to find them able to quote members of the medical research community as saying, even informally, that '99 per cent of research turns out to be useless',[3] or describing as 'plain nonsense' the idea 'that fundamental truths are revealed in laboratory experiments on lower animals and are then applied to the problems of the sick patient'[4] (shades of penicillin!). The former type of remark is always by hindsight, and never specifies whose particular research it is that is so lacking. The latter reflects some of that familiar strain between headquarters and the trenches, preclinical laboratory and clinical practice, theoretical physicist and materials man. Fortunately it is diminishing as fields of research increasingly overlap.

But a deeper doubt has been expressed by such as Sir William Osler (1849–1919). He was a great exponent of pathology, a new and exciting subject in his day; he contrasted 'pills and potions' with the 'gains of science' that the pathologists of the day were achieving.[5] This was before the therapeutic revolution from 1930 onwards, and his remarks about 'quack-like promises to heal' and 'continuate and inexorable maladies', quoted earlier, would have seemed reasonable comment. Bacon's 110th aphorism is really astonishingly apt; just so did Osler, possibly the greatest physician of his time, find it incredible that a 'great group of maladies' be other than permanently incurable.

Bacon's second comment, on discovery becoming taken for granted, is reflected in the other remarks. At one moment, we just cannot tell what is going to be really important; a little later, we know the answer, forget the rest, and take for granted all the earlier useful negatives, the exclusions of unnecessary further work. It is a familiar picture. After an Olympic games, it seems that only a few athletes ever achieve anything significant; but athletic clubs all over the country (from whom the winners came) know better.

It is not pessimism, nor retrospective disparagement, that hold the key to the future, but hope and imagination.

## Summary

1. The argument in this chapter turns to our expectation of future benefit. We see, first, a sort of predictable unpredictability. While we could well expect that work on an anaesthetic or on transfusion technique or on surgery of the heart would lead to an obvious clinical application, we could not expect that a physicist would generate a standard medical diagnostic tool or that a study of a disease in cattle would be the source of an important anticoagulant for man or that the noticing of an effect of a drug for treatment of infection would lead to a diuretic and better blood-pressure control. The greater the originality, the less the outcome or its timing can be foreseen.

2. We have to recognize the danger of being preoccupied with benefit rather than knowledge. Knowledge has a kind of inevitable priority, and our confidence in the expected benefit depends on the soundness of the knowledge.

3. Because the body is so complex and there is so much we do not know, work with the intact organism remains important; it is a superb detector of the unforeseen.

4. The very individuality of men and animals creates variability

of response, and hence the need to make not one, but a number of observations. The precision required from a result inevitably has to be 'traded off' against numbers of animals used; and nobody seriously interested in animal welfare can afford to be unfamiliar with the statistical techniques developed by biomedical statisticians to make that balance the best possible.

# 7

# Pain, suffering, and loss of animal life

When considering the costs incurred by the advances in knowledge and practical benefit involving animals, the chief issue in the past was the infliction of pain, and we have already outlined some of its physiology (pp. 31–4). Today concentration on pain alone is justly regarded as too narrow, and we need to take account of less tangible experiences: 'suffering', 'misery', and 'distress'. In addition, but also as a principle arising from the concept of a 'right to life', loss of life or disturbance of normal way of life may also be issues. With all these, great difficulties of definition arise, and Sir Thomas Lewis's dictum about pain is worth recalling again: that it 'cannot be defined, but is known by experience and illustrated by example.'[1]

## A possible categorization of pain

Yet, despite all the difficulties, it seems worth attempting some degree of categorization—first of pain—if only to fix our ideas. First, we can readily accept that everyday life for men and animals alike cannot be totally free of discomfort. Whether in exploration or play or with a posture held too long or with minor disease or trauma, there is a certain level of discomfort which is unavoidable, which no one would propose to treat and which may produce some minor movement but essentially does not interfere with normal activity. These could be termed 'everyday' discomforts. In human life we can recognize a next higher level, where normal activity can continue, possibly with some effort, and where to take an aspirin or an equivalent pain-killer can seem appropriate. Examples would be headache, menstrual pains, 'rheumatism', a pulled muscle, or a sprain. There is a characteristic response in animals that seems analogous, the rather misleadingly entitled 'writhing' test in mice. Here some mild irritant, that would not be felt on intact skin but would on an abrasion, is injected into the peritoneal cavity (the lining of the abdomen). The response is a posture as though the animal was trying to rub something away from its abdominal wall, with episodes of a characteristic movement and posture of its hind legs. But while it is doing this, it can still be seen to be exploring any novelty in its environment, head turning and whiskers twitching, in an entirely normal way. The response is of interest because it was the first animal test that was found to be sensitive to aspirin. Perhaps, therefore, we can characterize such pains, which modify but do not seriously interfere with activity and are treatable at the aspirin level, as 'minor'. The third category is 'severe' pain, where the pain is overriding and other than momentary, so that activity tends to centre round it while it is experienced or liable to be experienced. It is for this that the opiates are needed; they are effective in animals as well as man. Examples in man would be a coronary thrombosis, a fractured limb, some cancer pains, some neurological pains, and some pains after surgical operation. A subdivision of this kind is included in the 1986 Act's Guidelines.

## Suffering

If pain presents problems of definition, suffering, misery, and distress are even less definite. The best we can do, it seems, is to extrapolate from human experience: post-influenzal misery, gastro-enteritis, and chronic infection form one category. Fear, anxiety, and frustration form another. Another term, which is widely used, is 'stress'; but it has become almost meaningless unless an illustrative example is quoted. The term became more common when it was adopted by Hans Selye in his development of the concept of a 'chronic adaptive syndrome'. He assembled evidence that while 'stress' might have many different causes— such as injury, disease, or exposure to extremes of temperature— yet a common response could be characterized. It could begin with the general emergency reaction, involving activation of the sympathetic nervous system and release of adrenaline and noradrenaline; it would then go on to release of a hormone from the pituitary gland, ACTH, which in turn stimulates the adrenal gland to produce the shock-preventing 'corticoids' such as hydrocortisone. If this adaptive response was not sufficient and the stress continued, a final stage of 'exhaustion' would be reached with impairment of many body faculties involved in growth, reproduction, resistance to disease, and general activity. One particular result of this approach was that it suggested that one might use the level of adrenaline or ACTH or hydrocortisone in a body fluid as objective measures of stress. This could be extended, on the assumption that the situations producing stress are those that produce suffering, to using these objective tests as a measure of suffering.

## Assessment of pain and suffering

The problem of knowing how much pain and suffering an animal experiences, whether in experimental work, in nature, in animal husbandry or veterinary practice, in culling wildlife in game parks, or in domestication, seems at first wholly intractable. Marion Dawkins, however, in a book entitled *Animal Suffering*, has performed a most useful service in bringing together the sort of evidence that can be brought to bear; and it is worth summarizing some of the possibilities.[2]

First, we may consider how far the animal's life has been changed from its natural course and the implications of this. For the animal experimenter this is particularly applicable to the general care of animals bred and housed for experimental purposes and for long-term experiment. It would be wrong to assume that natural life is ideal, and that any deviation from it is necessarily adverse. Presumably domestication of an animal must be regarded as an unnatural state; yet to release a domestic animal into natural life in the wild could well lead to its rapid death at the hands of predators. The study of 'imprinting' has shown how critical are conditions in very early life, so that the conditions of an animal's breeding may be the best pointers to its needs in later life. While animal freedom is evidently desirable in general, it is not always clear that animals necessarily suffer if they cannot behave as they would in the wild. Indeed, human protection could make them better off, for natural life cannot be regarded as free from suffering: witness the high mortality, for instance, of a songbird in the wild (1–2 years), compared with that in captivity (11 years or so).

Second, we may look at the general health of the animal: its eating and drinking, growth, physical appearance, and reproductive performance. Of course, a particular index such as favourable growth is not, in itself, decisive; putting on weight might be simply an index of excessive confinement. Similarly, lack of any overt effect on health does not exclude suffering; if the suffering is brief, there may be no time for evident health defects to appear.

Third, it may be possible to establish physiological measures that can be taken as signs of suffering. But these are not in themselves decisive. There are considerable uncertainties in correlating actual levels of adrenaline or adrenal corticoid levels with degrees of suffering, both because of the variability between individuals and because of all the other, higher influences that affect suffering (such as memory, expectation, distraction). In addition, release of these hormones can be regarded as part of the general homeostatic regulatory mechanism of the body, and their activation does not always reach consciousness.

Fourth, we may look at the animal's behaviour, not only obvious avoidance or reflex responses, but other changes, such as change in the grooming of fur, or signs of conflict, apathy, unnatural aggression, or abnormal biting or other stereotyped pattern.

Fifth, it may be possible to determine an animal's own preferences by the choices it makes. This connects with the techniques used extensively in experimental animal psychology, whereby animals are trained, either by an aversive stimulus or by some reward (such as food or drink) to perform certain actions. If one assumes that what it chooses is in fact preferred and that what it avoids would involve suffering in some sense, then one is getting direct information. Experiments of this sort showed for instance that, contrary to human expectation, hens prefer a fine mesh flooring to a stronger, wider gauge metal flooring. There are, of course, difficulties. It would be very hard for man to anticipate all the factors influencing animal choice. It is also far from clear that animals would choose what was for their own benefit (the sight of a cat trying to remove the dressing over an infected eyelid is sufficient example). Yet there remains an area where useful evidence could be obtained.

Finally, we can extrapolate from human experience, although the dangers of doing this are obvious. Medawar cites the charming example of a little girl who thought a frog in the garden needed warming up, because it was cold.[3] Yet, if combined with sufficient appreciation of the animal's normal physiology and life-style, particularly when we are considering other mammals, such arguments from analogy seem legitimate.

There are, therefore, various approaches, all potentially informative, though none decisive in itself. It falls to those concerned with animal welfare to piece this evidence together. In doing so, experience is at a premium. It requires training even to know how to pick up an animal without causing it distress. It is a field in which the veterinary profession can give excellent advice.

## The development of skill in analgesia

We have possessed efficient analgesics ('pain-killers') for many years. The prototype of the powerful analgesic is morphine, the main active ingredient in opium, which has been known for centuries. Of the milder analgesics the prototype is aspirin, discovered in 1899. But both drugs have other actions which complicate their use in relieving either human or animal pain. Morphine, as well as being liable to produce craving and tolerance if given repeatedly and then withdrawal symptoms if subsequently withheld, also constipates through its action on the bowel's movements, depresses the breathing, affects the circulation, which can cause fainting, and may produce a reaction akin to allergy by direct release of histamine in the body. Aspirin is commonly an irritant in the stomach, and in large continued doses can damage the kidneys, as well as producing allergic reactions in some subjects. Pharmacological research and industrial development have done much to widen our choice of drugs, producing alternatives that are active by mouth or injection and have varying potencies and addiction liability, varying duration of action, and various degrees of associated action. The hospice movement, the development of pain clinics, and work by anaesthetists have all simultaneously benefited by and stimulated research into the control of pain, and many useful ways of combining analgesics with tranquillizers or other drugs have been found. It is probably now true to say that, in principle, no one need suffer prolonged severe pain, even if for certain conditions some discomfort is bound to remain. The essential problem in human practice is how to make available the knowledge and skill required on a wide enough scale.

If we now turn to pain in animal experiment, it is clear that *in principle* a good deal could be done to mitigate it. But two major difficulties must be recognized. First, much of our detailed knowledge of pain control is limited to human experience; and it by no means follows that the details of analgesic treatment will be identical in animals, or even the same from one animal to another. In particular the duration of action of a given dose may be considerably different. Secondly, the associated actions of the analgesic could well distort the results of the experiment. In turn this could well necessitate an increase in the number of experiments done, because extra experiments would be required to disentangle the effects of the analgesic from those of the test procedure.

In recent years the RSPCA has reviewed the scientific literature for experiments in which pain and suffering are evidently involved. Other bodies have made similar studies from time to time. In general, severe pain is rarely, if ever, produced. But we can tentatively identify some areas where it seems likely that the use of suitable drugs offers scope for reducing pain and suffering: in post-operative care, in research on infections and their treatment, in cancer research where tumour growth is substantial, and in tests where chronic exposure to a drug is necessary and where the drug produces some adverse effect.

It would be useless to insist, at once, that in such instances analgesics must now be used. Adequate knowledge of how to allow for the presence of the analgesic does not exist, so that many more experiments would be required; and it is not at all clear whether there is yet sufficient knowledge of potency and duration of action of suitable drugs for their safe use in all the species involved. In recent years, the concept of an 'institute' for the development of alternative methods has been forcefully advanced. The reader may feel, however, after reading the account of the development and scale of use of alternatives already in place (pp. 132–42), that such an 'institute' would hardly be a useful investment. Instead it would be more useful to set up the deliberate funding, at suitable centres, of research into methods of relief of pain and suffering in animals, particularly in the context of experimental work. To some extent it would be rather severely *ad hoc*: to discover, for instance, the most specific analgesic, with fewest side-effects, lasting the right length of time for the relief of post-operative discomfort in rat, rabbit, cat, or dog; or a drug which relieves any suffering that arises in cancer research without distorting the process of tumour growth or of tumour destruction by some test drug. Systematic knowledge of potency and duration of action for a range of drugs in a range of species would itself be of great use, particularly if associated side-effects could be delineated. Similar considerations apply to the use of local anaesthetics which can be used when a source of pain is strictly localized to the area supplied by some identifiable sensory nerve. Such work would, however, be likely also to yield some more general benefit. There has been considerable improvement in our understanding of the mechanisms of the production, recognition, and control of pain; and it is likely that the experiments could be done in such a way as to exploit and further advance that understanding. Further, one must remember that, while most experiments are done on normal healthy animals or people, in fact both humans and animals when they receive drug treatment are often diseased or receiving other drugs. To test an anti-cancer drug on an animal receiving an analgesic may well be a more realistic model of treatment in man, even if it also becomes more complex.

It is with work of this sort in mind that the idea of categorizing pain was outlined earlier. It would be foolish to use a sledge-hammer to crack a nut, and one of the tasks would be to develop pain-relieving regimes appropriate to the nature of the suffering involved.

As with the assessment of pain, this is a field where veterinary expertise could well be especially helpful. Realistic appraisal of suffering is obligatory; and without adequate experience, a layman might, for instance, propose post-operative analgesic drug treatment for a rat in whom a recording device had been implanted under anaesthesia, which might impair the return of eating, drinking, and normal activity, and actually prolong rather than mitigate post-operative distress.

## Summary

1. Since it is necessary to assess in some degree the pain that may be incurred both by animal experiment and by the failure to do

animal experiment, a possible categorization of pain is proposed:
(a) 'everyday' pains that do not interfere with normal activity and
are accepted as not worth treating; (b) 'minor' pains, that may
modify but do not dominate activity and are responsive to aspirin-
like drugs; (c) 'severe' pains, where pain is overriding, calling
for opiate drugs. It is agreed that categorization offers great difficulties; yet evidently some pains are worse than others, and
'bench-marks' could be helpful. There seems no way at present to
categorize suffering or distress.

2. While pain and suffering in another being are, strictly
speaking, unknowable, yet for practical purposes there are a
number of lines of evidence of their presence and intensity which
deserve attention: (a) disturbance of normal pattern of life; (b)
impairment of general health; (c) physiological indices of 'stress';
(d) changes in behaviour patterns; (e) expression of an animal's
own preference; (f) extrapolation from human experience. No
single index is reliable; but attention to the various aspects, combined with experience with animals, can improve the assessment.

3. It is suggested that more research could usefully be done to
discover suitable procedures for minimizing any pain and suffering
involved in experimental work. The two main difficulties are that
present knowledge of analgesic techniques is largely based on
human trial, and that the introduction of analgesics into an experimental procedure could, by complicating it, increase the number
of experiments required. Deliberate research for the development
of analgesic procedures of the highest specificity and suitable for
different species could reduce these difficulties.

4. Veterinary expertise should be deployed to assist in the
further development of skills in anaesthesia.

# 8

# The alternatives to animal experiment

It may be accepted that animal experiment has been justified in the past and will be in the future, yet the concern remains that *unnecessary* experiments not be done. We must turn, therefore, to ways in which this might be avoided.

## Avoidance of unnecessary repetition

One charge is that the same animal experiments are frequently repeated, for no good reason, and that considerable savings could be made by stopping this. There are some obvious points to make. First, some repetition is absolutely necessary: without it a result can easily be dismissed as 'unconfirmed'. It is also essential, if others are to build on previous work, that they establish a genuine link with it, which involves, at the start at least, similar experiments. Often, too, control experiments of some sort are needed, to be done under the same conditions as some test procedure, and these may well appear repetitive. One must recognize, also, the forces working against such unnecessary repetition: cost, waste of time, and the existence of many people, such as heads of departments, other scientists, and grant-giving bodies, only too willing to point these out. But the question can still be asked, whether such redundancy occurs in practice or not. A short answer is that the Littlewood Committee (1965), in its very thorough study for the Home Office of British practice in experiments on animals, considered this point specifically, took evidence about it, and concluded that the risk of unnecessary repetition of experiments was small and the scale of duplication not serious.[1] Their discussion of the whole matter, about which evidence was widely taken, is well worth reading.

Since then, the literature has of course expanded very greatly, so the position might have deteriorated. But computer aids have now also become far more widely available and much more powerful. Should it not now be possible with such aids to avoid any repetition at all? The question of the efficiency of information retrieval was the subject of a valuable study in which 14 bodies (2 university, 1 governmental, and 11 industry) took part.[2] It concerned particularly the recovery of information in chemical toxicology and involved framing eight test queries, designed to represent the sort of information likely to be sought, sometimes very specific, sometimes rather general. The various participants then conducted a search, using the sources available that seemed most appropriate and using methods that would be used in real life (such as following up references and browsing through a wide variety of handbooks and textbooks, printed indexes, on-line data bases and data banks, and in-house material). The results and techniques used were then compared, and a consolidated list of references obtained, from which the value of various sources and methods of retrieval could be estimated. Some difficulties very familiar to the scientist were highlighted.

First, with one query, most of the literature pre-dated modern abstracts and data banks, and was found only in old textbooks or individual papers. This difficulty will always be present to some degree, even as the methods of data storage and retrieval steadily improve; it will never be possible to bring *all* past information into the format used in the latest data-storage system.

Second, the required information could exist as a minor item, not indexed, as part of a large paper on some other topic.

Third, the on-line type of retrieval, where a large list of brief titles can be obtained, was sometimes effective; but it sometimes failed to allow a desirable *evaluation* of the reference in the way that personal search may do, and could lead to a great excess of irrelevant material.

Fourth, there is a major dilemma in obtaining a balance between asking too general a question (leading to flooding with unwanted material) and making it too detailed (when specific details are liable not to be indexed, or not indexed in that precise form).

Finally, following on the above, it became apparent that literature search, as well as the selection of the specific material to be indexed, have both become highly specialized tasks. If this specialization develops too far, however, it would put effective search beyond the reach of all save major centres.

At the same time it seemed clear that there were two or three rather effective and commonly used principal sources of information which, taken in conjunction, could produce fairly readily up to two-thirds of what appeared to be the total literature available. This is rather reassuring. For if the question is 'Has the experiment I propose to do been done before?', the chance of this is in any case small; and the information successfully retrieved is very likely to settle the major questions: namely, whether there has been *any* prior study and, if so, the general properties of the substance in question. It needs to be realized that such searches will be conducted *in the light of knowledge at the time of the search*, which will be different from the earlier state of knowledge. This is the reason why precise reduplication of experiments is in fact uncommon, since the questions to be asked change as knowledge increases.

## Use of the non-animal alternative

Historically, the interest in non-animal methods seems to have arisen in response to the experimenter's claim that physiological discoveries could not have been made by any other means. The challenge was met by suggesting other ways in which the discoveries could have been made. For a long time, the alternatives suggested were chiefly 'mortisection' (in other words, dissection of dead humans or animals), clinical observation, and a priori thought—the methods used over past centuries, whose success, but also whose failures, are obvious enough. But the advance of biomedical science steadily made more analytic procedures possible, using chemical, biochemical, tissue culture, or other techniques; and their evident contribution to knowledge in turn suggested that yet more could be expected of them.

The idea of 'alternative methods' as a distinct entity crystallized in the 1960s, and came to public prominence chiefly, perhaps, with the Bill brought forward in the House of Commons by (the then) Mr Douglas Houghton in November 1972. This sought to amend the 1876 Act by making it an offence to do an experiment on animals which could be done by alternative means not involving an experiment on a living animal. Although there was widespread agreement that alternative methods should be encouraged, the Bill was talked out at its third reading. One of the particular points made in debate was that it was sometimes very uncertain whether a method *was* alternative or not. A striking case was cited in which a drug firm wished to use a chemical method for standardizing a pituitary hormone used in medicine; after two years of debate, however, the British Pharmacopoeia Commission insisted that biological tests should still be used, because of certain chemical uncertainties. Also, despite wishing to encourage such alternative methods, it seemed unreasonable to make it a penal offence for an individual investigator not to have discovered, in the whole vast literature of medical science, that an alternative method existed. Finally, some measurements—for instance, of blood adrenaline levels—could indeed be made by chemical methods, but only at considerable expense compared to tests on an anaesthetized rat. The effect of the new legislation would have been to penalize severely the smaller centres of work. It fell to the Research Defence Society, with a budget less than 1 per cent of the welfare societies, to draw attention to these difficulties. Subsequently there has been considerable discussion and a number of substantial meetings and publications on the subject, which have clarified the position. An excellent book by the late Professor D. H. Smyth, commissioned by the Research Defence Society, gives a full and independent discussion of the issues involved, after very wide consultation with all interested parties, together with much of the evidence available in 1978.[3] Since then, the main developments have been in toxicity testing (see Chapter 10).

For the research worker himself, it seems strange that he should find himself accused of neglecting alternative methods or, indeed, that they should be regarded as something new to take account of; for it is the animal experimenters who have themselves been developing them and using them for decades. Rather than repeat the accounts of the limitations of such methods (see pp. 115–17), it is worth explaining their development and use.

There is a natural path that continually opens up before the experimenter as opportunities for deeper analysis on simpler systems become possible as a result of knowledge gained in animal work. For instance, with a drug like digitalis the first experiments were on whole animals, since it was not known on what part of the

**Table 8.** Methods of experiment not involving whole animals

| Isolated perfused organs | Isolated tissue or tissue sample | Isolated single cell | Subcellular constituents |
|---|---|---|---|
| Liver | Striated muscle | Fat cells | Nuclei |
| Muscle | Electroplax | Liver cells | Mitochondria |
| Heart | Iris | Neurones | Microsomes |
| Lung | Trachea | Glia | Lysosomes |
| Adrenal | Bronchi | Muscle | Synaptosomes |
| Intestine | Lung | Smooth | Cell membranes |
| Skin | Intestine | muscle | Actin/myosin of |
| Spleen | Spleen | Red cells | muscle |
| Kidney | Uterus | Leucocytes | |
| | Seminal vesicle | Platelets | |
| | Vas deferens | Mast cells | |
| | Bladder | | |
| | Salivary gland | | |
| | Fat pads | | |
| | Liver slice | | |

body (brain, heart, blood vessels, alimentary tract, or kidney) it worked. Once the site of action is identified—such as the heart with digitalis—then work on an isolated, perfused organ becomes possible. After that may come the use of isolated tissues, then single cells, then subcellular constituents. The development depends on the previous animal work. The conclusions drawn from the more analytic work also need to be checked in the whole animal to establish that what happens to pieces of cells in a test tube also happens in the intact tissue. All the alternatives I know of have traced this path.

Tables 8 and 9 list those methods that can be called in some sense 'alternative' that are familiar to me from the literature and from my own work. The word is used in various ways to mean either avoiding experiments on a living animal (although animals will have to be killed to obtain the material) or avoiding both live experiment and the killing of any animal (Table 8), or, at the extreme, avoiding the use of living matter altogether (Table 9).

Experiments with living isolated nerve and muscle began with Galvani in 1791, and continued through the nineteenth century. The first attempts at studying the heart in isolation go back at least

**Table 9.** Procedures not requiring living matter

| Structure–activity relationships | Analogue or computer modelling |
| --- | --- |
| Curare (1868) | Cardiovascular function |
| Nitrites (before 1900) | Anaesthetic uptake |
| Anaesthetics (1899) | Decompression sickness |
| Sympathetic amines (1906) | Drug metabolism and distribution |
| Barbiturates (before 1930) | Receptor function |
| Sulphonamides (1940) | Neuromuscular control |
| and many others | Nerve action potential |
| | Control of respiration |

to 1846. But success depended on knowledge of the essential constituents of the supporting fluid. This came only in 1880, when a lazy laboratory assistant, using tap water instead of distilled water in preparing solutions for use with a frog heart, allowed Sydney Ringer to recognize the need for calcium ions. That was the beginning of a trail in which control of acidity, glucose, other salts, and many other factors were gradually recognized, leading up to the tissue culture medium of today. Early attempts, dating from around 1900, were also made to perfuse isolated organs with blood, but initially could be attempted, for lack of any non-toxic anticoagulant, only by using blood that was allowed to clot but with the clot removed as it formed; an organ could then be perfused with this 'defibrinated' blood, which retained, of course, many of its normal properties. This was only partly satisfactory, since the process of clotting released pharmacologically active substances which could disturb the blood flow. Yet this, too, was a fruitful observation; for the attempt to characterize one of these substances led ultimately to the identification of 5-hydroxytryptamine (5-HT), contained in the platelets of the blood; and 5-HT in turn was later revealed as one of the amines of profound importance in brain function. (A further by-product is work using the blood platelet, whose diameter is less than one-twenty-thousandth of an inch and of which we possess about a million million, as a 'model' and 'alternative' for the brain!) Another fascinating experiment found that to pass the defibrinated blood through isolated lungs 'cleaned it up'—an early indication of what is now recognized as an extensive metabolic capacity of the lung. Subsequently, reliable and safe anticoagulants—some occurring naturally in the body, like heparin, others made artificially—were discovered, and now play their part in the heart-lung and kidney machines of modern surgical practice which themselves evolved from the early perfusion experiments. As a result of these developments, most organs and glands in the body have now been studied in isolation either (if thin enough to allow sufficient oxygen to reach the depths of the tissue) by suspension in a suitable fluid or by perfusion. These experiments do not avoid loss of animal life, but do avoid experiment on the whole living animal. Some preparations, such as the stimulated intestinal strip, have been remarkably useful, because they possess two nerve networks (one controlling propulsive activity, the other probably secretion) as well as muscle and a wide range of pharmacological receptors; and a dozen preparations can be obtained from a single intestinal length, allowing considerable economy of animal life.[4] This was among the alternatives selected for illustration in a European Medical Research Councils survey.[5] The sensitivity of the preparation to morphine-type (pain-killing) action, amongst other actions, enabled it to serve as a screen for this type of activity, helping the discovery of the endogenous opiates and saving much *in vivo* testing of pain-relievers. It is one of the more successful alternatives, although rarely cited as such.

The next analytic step was tissue culture, inaugurated by Harrison's classical work culturing nerve cells in 1906. As with isolated organs and tissues, success has grown with knowledge of the essential constituents; and tissue culture is now a routine technique. In a parallel development, special tools were designed for manipulating tissues under the microscope; methods of electrical recording of great speed and sensitivity were discovered; and other physical, chemical, and biochemical techniques were developed so that experiments could be made directly on single cells a fraction of a millimetre in diameter.

The same analytic drive led to what is called 'subcellular fractionation', when, around the 1950s, ways were found whereby cells themselves could be gently broken and the various 'organelles' within them separated from each other. These included 'mitochondria', little factories producing energy from our food in a form the cell can use; 'lysosomes', which degrade unwanted materials; 'microsomes', which convert foreign substances to less harmful and more readily eliminated forms; 'synaptosomes', containing the junction between two nerve cells, with the chemicals and receptor apparatus together; and cell membranes themselves, with all their machinery for responding to chemical and other stimuli, for transporting materials in and out, for secretion, and for recognizing foreign substances and other cells. From muscles, the proteins whose relative movement results in the shortening of muscles were isolated (including actin and myosin), so that simple *in vitro* systems could be studied. Today, inspection of the literature shows that experimental research conducted on such isolated organs, tissues, cells, or cell constituents has grown steadily over recent decades, and there are now more experiments of this sort than there are experiments on whole animals.

The account thus far of the non-animal alternatives points to their successful aspects. But the earlier discussion (pp. 115–17) of the complexity of the whole body and the continuing necessity for animal experiment must still be borne in mind. Two illustrative examples of the dangers of experiment on other than the whole animal are worth citing here.[6] The first concerns the discovery of the sulphonamides. P. Domagk's original experiment used a red dye, Prontosil rubrum, which he tested against a streptococcal infection in mice. It was also found that it was inactive when tested *in vitro*, which provided a puzzle until it was shown that Prontosil owed its activity to being broken down in the body to produce the active material, sulphanilamide. It is singularly fortunate that Domagk included a test in an animal. That was in 1935. Nearly 50 years later the same happened with an important new antiparasitic agent, active against roundworms (nematodes); this proved inactive by *in vitro* tests, but given by mouth to parasitized mice had a quite astonishing potency, being active with a dose of less than one millionth of a gram.[7] Perhaps the essential point is that it is wrong to view animal and non-animal methods as opposed to or competing with each other; in fact, they serve different but complementary purposes.

## Prediction of biological action from chemical structure: structure—activity relationships

Two other approaches also have a long history. Today's use of modern chemistry (based on quantum mechanics) to predict pharmacological action reaches back to 1868, when the first great generalization in the relation of chemical structure to action in the body was made by two Edinburgh scientists, Alexander Crum Brown and T. R. Fraser. They found that a particular pattern of combination of carbon and nitrogen atoms ('quaternary nitrogen') led to paralysis of the voluntary muscles similar to that produced by curare. A second great step was in 1900, when another generalization was made, that any substance liable to dissolve in fats to a given extent would be an anaesthetic. These two generalizations still guide us today, and were followed by a growing stream of similar generalizations as chemistry and pharmacology advanced. The great virtue of modern quantum chemistry is to give a far more realistic picture of how a chemical substance 'looks' to its receptive site than is provided by a chemical formula drawn on paper; and while the calculations are still imperfect and are so laborious that they are only practicable with computer aids, they already provide a useful predictive capacity in certain fields. The strategic difficulty remains, of course, that their power is only as good as the biological information on which they are built. The chemical structures studied have always been identified as significant by other, biological means.

## Modelling

The modelling of biological phenomena is another area of interest. This too has a long history. The practical class offered to Cambridge medical students in 1876 included experiments with rubber tubes, glass connections, a pressure-recording device, and a hand pump, to teach the principles of the circulation.[8] This has been succeeded by more sophisticated models, using analogue or digital computers to illustrate, say, the principles governing the rise and fall of the amount of anaesthetic in the blood and the brain if ether, chloroform, or nitrous oxide (laughing gas) are inhaled; or how the action potential arises and travels along a nerve or muscle when they are excited into activity; or how the centre in the brain controlling our breathing may switch alternately between breathing in and breathing out; or the mechanism by which our eyes or hands track a moving object. But it is important not to exaggerate the usefulness of such models for teaching. Anyone who has used them (or examined students subsequently) comes to realize that unless the model is very simple and vivid, too much time necessarily gets spent on explaining the analogy and justifying its validity, at the expense of time devoted to the actual biological principles to be demonstrated. Such models also convey a rigid mechanical impression of biological responsiveness, a far cry from the individual and infinitely variable pattern of real life. The importance of such modelling can, however, be substantial for research purposes—essentially for exploring the results of some theory over a range of variables. But here, too, the models are only as good as the biological data fed into them. As is so true of statistics also, 'Garbage in, garbage out'.

## Alternatives: a reductionist philosophy

As one reviews the methods proposed for replacing animal experiment, the reductionist philosophical background becomes steadily more apparent. The object of biomedical research is to improve the understanding of the human and animal body and its functioning. To do this, it has often been helpful, for purposes of analysis, to use isolated organs, isolated tissues, and cells; but this has always been to allow deeper analysis, and awareness of the problem of integrating into a functioning whole has remained. Indeed, there is a trend today, as biological knowledge has grown, for thinking to move in the other direction. This comes partly from the need to put pieces together again. But a more important reason is that processes must be studied at the level of complexity at which they occur. The laws of physics and chemistry are capable of an infinite variety of exemplifications; the question is, which particular exemplification are we dealing with at a given moment, and of what type? It is particularly clear in sociological research, where no one expects to solve the problems in terms of electronic orbitals. Likewise, one must deal with the whole organism in tackling, say, epilepsy or schizophrenia. The nature of the answers, when one comes to them, is necessarily in terms of higher-order functions. So the scientist moves between the reductionist and integrative poles, as the problem demands.

But the alternatives movement is based not on scientific considerations, but on a particular view about human and animal welfare. It claims that, ultimately, sufficient information can be obtained from isolated parts of the organism, from cell culture, or from models and computations about them, to make possible total replacement of work on the whole animal and total reliance on these fragments to represent the whole creature, both for purposes of safety regulation and for the advance of knowledge. This should be recognized as one of the most reductionist enterprises in existence, and entirely incompatible with any concept of integrative functioning of whole organisms.

## Are alternatives sufficiently used?

Despite the history of the use of these methods, it may still be thought, however, that there is insufficient pressure on the animal experimenter to employ them. It is impossible to respond to such a comment other than by pointing to the pressures already in existence and to evidence that the methods are in fact used. The pressures for their use are their scientific usefulness, their relative cheapness (usually, though not always), the ability to generate more standardized testing procedures (of especial importance in industry), the formal requirement by the Home Office and bodies such as the Medical Research Council to verify that an animal experiment cannot be replaced by one avoiding animal use, and, not to be forgotten, the feelings of humanity, which the animal experimenter has no less than other people. Objective evidence for the progressive use of such methods as they have become available can be found in the Home Office Annual Return, which gives the number of experiments and the number of licensees.[9] From this it can be calculated that over the 15 years between 1968 and 1982 the number of animal experiments per licensee has fallen by 50 per cent from 405 to 203. Yet the output of biomedical research by any test has greatly increased.

A second interesting calculation is a comparison of animal use with research expenditure, shown in Table 10, covering the last 20 years. One may take as an indicator of general medical research the expenditure by the British Medical Research Council, whose support extends across the whole spectrum from the molecular to the clinical.[10] That expenditure needs correcting for inflation; and the third line of Table 10 gives the MRC budget in 'real' terms, namely 'January 1974 £s'. The fourth line gives the number of animals used over the period in Great Britain. The last line then takes this expenditure as an index of national biomedical research

**Table 10.** Comparison of animal use with research expenditure

|  | 1962–3 | 1964–5 | 1967–8 | 1971–2 | 1978–9 | 1981–2 | 1990–1 |
|---|---|---|---|---|---|---|---|
| MRC budget (£ million) | 6.27 | 9.63 | 15.1 | 25.1 | 61.8 | 106.5 | 204.3 |
| Retail price index (Jan. 1974 = 100) | 53.5 | 57.1 | 63.4 | 82.9 | 207.2 | 277.3 | 476 |
| MRC budget (Jan. 1974 £s) | 12 | 16 | 24 | 30 | 30 | 38 | 43 |
| Animals used in UK (millions) | 4.2 | 4.8 | 5.2 | 5.3 | 4.7 | 4.2 | 3.2 |
| Animals used nationally per MRC 1974 £ | 0.35 | 0.30 | 0.22 | 0.18 | 0.16 | 0.11 | 0.07 |

activity, which is, if anything, conservative in relation to industrial expenditure, and shows the animals used nationally per 'MRC 1974 £'. It will be seen that it has declined over the 20 years by about 80 per cent.

Among other ideas, there has been considerable pressure for the establishment of some institute dedicated to the development of alternative methods, if necessary by diversion of the requisite funds. It is very doubtful that this is really needed, when one thinks of the proliferation of laboratories dedicated to cell biology. It would be more sensible to let such methods evolve from the scientific context in which they are needed, rather than to try to develop them *in vacuo*, with no answer to the question 'Alternative to *what*?' But there is one strictly practical point, which arises very clearly in questions of standardization of drugs by biological test. With these, satisfactory biological methods have been worked out, and doctors or vets and regulatory authorities alike rely on the products tested in this way. If some non-animal method is introduced, it can be accepted only if it is at least as reliable as the previous method; and to prove that, an extensive comparison between them is necessary, involving animal experiment that would otherwise not have been done.[11] Care is therefore needed, for if the new procedure proves inadequate, all the extra animal experiment is wasted. In fact, both for humane and other reasons, there is considerable continuing interest in such moves, in both industry and research institutes. It is doubtful if, on the research side, much more could be done without incurring considerable additional animal use. Of much more interest, however, as is discussed later, is how far regulatory bodies can relax over-stringent standards, reducing animal use in that way.

## The pattern of growth of new remedies

Such findings may be regarded as rather encouraging. Yet those who are concerned that we do not lose or delay future medical advance may also wish to be assured that the advance of knowledge is not being hindered. It is not easy to assess this, but one approach is simply to review the rate of introduction of new remedies. This, indeed, is worth a digression, since it may help to place modern therapeutics in a wider context. If we review the old pharmacopoeias, we can recognize a variety of stages.[12] As we can see from Figure 15, the first pharmacopoeia in this country, prepared by the Royal College of Physicians in 1618, contained over 2,000 remedies. The initial stage was the pruning of the traditional herbal and other remedies—most of them both ineffective and innocuous, and some quite bizarre with scores of herbal and other constituents. By 1746 there were only about 650 items, not only fewer but much less complex. This history really represents a sieving out of man's experiments over the centuries with the herbal and other substances in his environment. The uncritical appeal to such remedies today often ignores those long centuries of unrecorded trial and error, by which the useless or harmful were shed, but which effective drugs like opium, digitalis, and belladonna easily survived.

The next stage may be taken as the first step towards obtaining the pure active principles, free of inactive or harmful dross. So we see the rise of the tincture and the extract (Figure 16), each providing a way to separate out the active principles according to physical properties—fat-solubility or water-solubility, respectively. This phase lasted a long time, even up to the 1930s. During this period the successive editions of the *Pharmacopoeia* of the Royal College of Physicians merged into the *British Pharmacopoeia* as we know it today, with its first edition in 1864.

Knowledge of microbial infection and of methods of disinfection, together with the introduction of the hypodermic syringe, made hypodermic injection of a few substances possible in the latter part of the nineteenth century; but injection did not 'take off' until chemical and pharmacological advance allowed the use

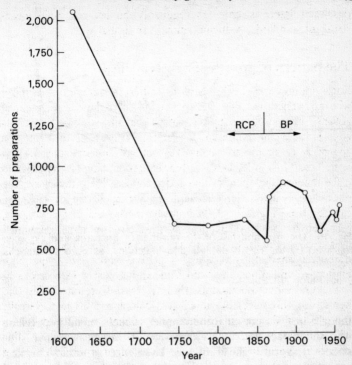

**Fig. 15.** The change in the number of the materia medica and derived preparations in the pharmacopoeias from the first pharmacopoeia of the Royal College of Physicians (RCP) in 1618 to the *British Pharmacopoeia* of 1953 (BP)

From Paton (1979b).

of pure active substances in the 1930s. This was catalysed by Henry Wellcome's insight into the virtues of a convenient form of medication of constant stable composition. The word 'tabloid', which he patented, introduced a new word into the English language. The era of pure substance, in injection and tablet form, began. The age of the tincture and the extract entered its last stages.

Two new factors now appear (Figure 17). The first, from 1932, was the use of biological tests, which were used to standardize those substances which still baffled the skill of the chemist. It was

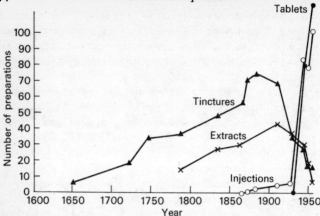

**Fig. 16.** The rise and fall of the number of tinctures and extracts, followed by the rise of tablets and injections, as pure substances became available, 1650–1950

From Paton (1979b).

thought in the 1950s that the number of these might be levelling off. But in fact the rate of discovery of biologically active substances still runs ahead of our knowledge of exact chemical constitution: that is, knowledge of 'activity' can still run ahead of knowledge of 'structure'. The other factor is the growth in application of synthetic chemistry. We have to be a little arbitrary about our definition of 'synthetic'; and the line drawn in Figure 17 to represent 'unnatural' drugs counts those not obtainable *as such* from natural sources (beginning with ether and chloroform). Most are wholly synthetic in origin, but with some a natural product is chemically converted to the drug we use. Each such chemical structure normally gives rise to several preparations, so that the number of these 'unnatural' substances by no means reflects the extent to which the *Pharmacopoeia* is now synthetic. (The word 'unnatural' is itself arbitrary: chemicals made by a plant and made by a chemist are all part of nature.)

All this traces the pattern of discovery of new remedies, but it does not give a picture of their impact. One way of estimating this is the rate of change of the *Pharmacopoeia* itself. This may be estimated by expressing the new additions to the *Pharmacopoeia*, at

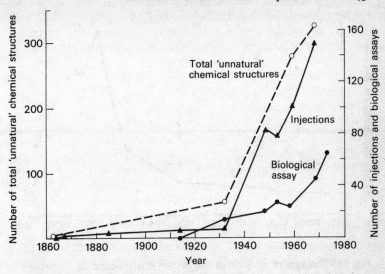

**Fig. 17.** The growth of the number of injections, of wholly or partly synthetic preparations, and of biological assay in the pharmacopoeias, 1860–1973

From Paton (1979b).

each revision, as a proportion of the total preparations. As well as additions, there are deletions, so that there is a sort of steady 'rinsing', which is in fact accompanied by quite a slow growth in total size of the *Pharmacopoeia*. Figure 18 shows the steady, fairly low rate of revision up to around 1930, then the accelerating growth up to 1963. At this point, there begins a clear and rapid decline, with the 1978 level of innovation roughly corresponding to that of 30 years earlier.

## The parallel rise and fall of pharmacopoeial innovation and animal experiment

It is in that continuing decline since 1963 that unease may be felt. The immediate cause was, of course, the thalidomide disaster, resulting in the UK in the formation of the Dunlop Committee, now succeeded by other bodies, which demanded much more extensive test procedures. But one might have expected that once

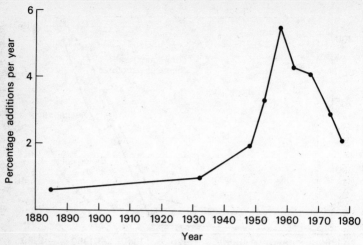

**Fig. 18.** The rate of revision of the *British Pharmacopoeia* expressed as the percentage new additions each year since the last revision From Paton (1979b).

new regulatory procedures had been established, growth might have started again, though from a lower level because of the greater cost of testing. Yet this has not happened. One possibility is that biomedical research is no longer sufficiently centred on integrated systems; as a result, too little research is now done in circumstances where there is the best chance that the unexpected can show us what we had not even imagined—that is, in experiment with the whole animal.

To explore this further, Figure 19 shows two graphs covering the last 60 years: one is of the number of experimental animals used each year under the Act, sampled since 1945 at 5-year intervals; the other is the figures for the rate of revision of the pharmacopoeia. The interval between pharmacopoeias has been variable, so the figures have been expressed as an annual percentage change averaged over the relevant period. A proportional (logarithmic) scale has been used, to help comparison; equal proportional changes in the numbers take equal distance. It should be noted that the animal returns are of totals, while the *British Pharmacopoeia* figures are of percentage rates of change.[13]

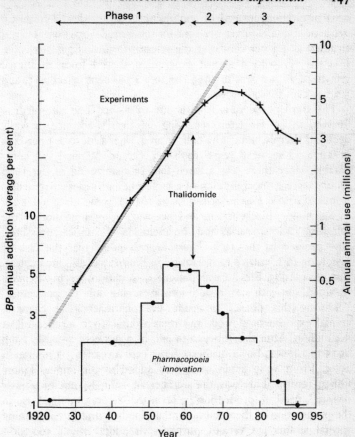

**Fig. 19.** The rise and fall, 1930–90, of the annual number of animal experiments and of the rate of revision (per cent annual addition) of the *British Pharmacopoeia*, both on a log scale

One can recognize three distinct phases. In the first, from 1930 for nearly 30 years, there was a roughly parallel rate of growth in both animal experiment and pharmacopoeial innovation, by 5–10-fold.

A second phase came after the thalidomide tragedy in 1961, with all its regulatory and other consequences. As noted earlier, drug innovation, instead of continuing to grow, slowed from 5.6 per cent annually to 5.2 per cent for the period 1958–63. Then it steadied at 4.3 per cent over the interval to the next *BP* in 1968. Likewise, growth in animal experiment slowed from an annual growth of 10 per cent in 1955–60 to 5.4 per cent in 1960–5, and 3.7 per cent in 1965–70.

The third phase runs from about 1968–70. The rate of drug innovation, so far from resuming its earlier advance, actually declined sharply, almost by half, from 4.3 per cent to 2.3 per cent in 1978. The next years 1978–83 cannot be analysed comparably, since there was a three-fold reorganization of the 1980 *Pharmacopoeia*: incorporation of a large number of lesser remedies from the *British Pharmacopoeia Codex* (*BPC*), some quite old, such as fig, honey, blackcurrant, and beeswax; accommodation to the *European Pharmacopoeia*; and the inclusion of surgical dressings. The preface to the 1980 *Pharmacopoeia* notes that the task of reviewing *BPC* entries for inclusion in the *BP* was still incomplete. But after 1983 the normal pattern of additions seems to have resumed, but with annual additions averaging only 1.4 per cent.

During this phase, alongside the pharmacopoeial changes, animal experiment slowed, and then around 1970 levelled off at 5.4 million. After this it began to fall, by 3 per cent annually from 1975 to 1980 and by a further 5.6 per cent a year to 3.2 million in 1985. From 1987 a new item was added to the annual Home Office return. This was the number of animals, not previously counted, used in 'production', a term that refers to animal use in the preparation of therapeutic sera, antibodies, and in maintaining special strains for various purposes. It added about 500,000–700,000 a year to the figures during the following 4 years. Discounting this, the rate of research animal use during 1985–90 fell further by about 3 per cent, to 2.7 million (3.2 million if 'production' is counted in).

Comparing these figures for 1970–90 with those for 1930–60, we see again a remarkable parallel: but now it is between the fall in animal use and the decline in pharmacopoeial revision.

This analysis strongly suggests that the decline in animal experiment has in fact already had a serious impact on the discovery of significant new drugs. This hypothesis must be considered.

## Measures of innovation

Discovery in medicine is an exceedingly difficult thing to quantify, because discoveries are so variable, and their significance is to some extent subjective. Pharmacopoeial revision was chosen to assess it as regards therapeutics for two reasons: that the *Pharmacopoeia* represents a sort of test not of prospective, but of *proved* clinical usefulness. As the preface to the 1885 *BP* puts it:

More (medicines) might have been added, but it was felt with most of them that they had not been sufficiently recognised by the medical profession, whilst in regard to others it was considered that there were already in the Pharmacopoeia agents having like properties and of equal if not of greater value.

If one studies the *pharmacopoeias* over the years, one sees older remedies being omitted (though still in use) while better new remedies take their place. The rate of deletion was quite high; up to 1973, deletions totalled over 1,000 and averaged about 70 per cent of the additions, so that by then the *Pharmacopoeia* had only grown to 1277, from about 900 monographs in 1885 (cf. Figure 15). The additions thus represent a genuine rinsing through with new proved materials. It is obviously possible that selection policy may have varied over the years, but there is no historical evidence to test for such fluctuations. It is hard to believe, however, that strong candidates for admission would ever be excluded for long, while no doubt reasons of economy would lead to the dropping of those items that had become less significant.

The second reason for the study of the *Pharmacopoeia* was that its preparation over the years has involved the consultation of a rather wide body of medical, pharmacological, and pharmaceutical opinion. While judgements of value are always difficult, the range and number of those who have been involved in framing the *Pharmacopoeia* have no rival.

The *BP* now totals around 2,100 monographs. The growth presumably reflects the widening range of therapeutics, as our understanding of disease processes deepens. The significance of any additions depends on what is already available. For this reason, it has been felt that the additions to the *Pharmacopoeia* should be expressed proportionately—i.e. as a percentage increase in the total *Pharmacopoeia*.

Two other tests of drug innovation can be considered. First

is the number of medicinal substances given 'approved names', regularly published in medical journals. When something interesting has been found and is beginning to be tested, it is convenient to have an agreed name for it. This requires the presentation to the authorities of a prima facie case that it will be significant. It would not be worth a manufacturer's trouble to ask for it unless there were novelty and promise. The system started in 1951 with around 35 names a year, rising to around 75 in 1962. Thalidomide caused a fall to 40 in 1973, since when, up to 1990, it has oscillated quite widely around an average of 56 ($+/-25$ s.d.). The variation is too large to reveal more than that significant recovery since thalidomide has not taken place.

A similar picture comes from a count of 'new chemical entities' introduced on to the market each year.[14] This ran at around 50 up to 1963, and since then has oscillated around 20.

A related set of figures, the 'number of licences for new active substances' granted by the Medicines Control Agency is available only since 1971, when the Medicines Act was implemented.[15] Setting aside the first 2 years of the licensing, the yearly average in four successive 4-year periods has run at 27, 26, 25, and 19, with a possible upturn to 38 and 36 in 1989 and 1990. The individual figures are very variable, ranging from 9 to 46. The lack of a consistent trend since 1973 is much like that of the approved names and the new chemical entities. It will be appreciated that these three names cover related but slightly different categories, which can all be regarded as 'candidate drugs'.

Thus the pattern is of a decline in candidate drugs after 1960 to an approximately steady level, but a *continuing* decline in proved effective innovation over the same period. In my view, that is precisely what one would expect; for the principal source of genuine new discovery can only be experimental work involving all the complexity and still undiscovered richness of the whole organism.

There is an interesting difference in the timing of the two peaks in Figure 19, which ties in with this. One would expect that after thalidomide there would be a new wariness about admitting new drugs to the *Pharmacopoeia*, so that additions would check at once. On the other hand, drug development programmes already fully committed would have to continue, but would need an increased use of animals to take account of new safety requirements. So it is not surprising that animal experiments did not immediately decline after thalidomide, but first slowed in growth, despite an immediate check in pharmacopoeial addition.

But a number of other explanations may be put forward:

(1) It could be suggested that the decline is related to economic conditions; but in fact there is no correlation with economic cycles.

(2) It is often pointed out by industry that the new regulatory demands impose a heavy burden on the cost of drug development. It is certainly true that development has become more expensive. But this has not been shown to have prevented drug development, and indeed the idea is contradicted by the rise in R&D expenditure: from around £35 million in 1970 to over £400 million by 1982. Nor has it been shown that these new costs are not recoverable from the user; one may expect the health services to be ready, on sufficient case stated, to pay for further assurance of safety; and their expenditure has certainly risen. It must be remembered that the early stages of significant discoveries are usually quite cheap; it is in finding the best particular drug in a group or series and in establishing its properties and safety that the costs arise.

(3) One might envisage some influence from the campaign for the use of 'alternatives' to animals. This could have contributed to a reduction in animal numbers; but these techniques were supposed to improve innovation, not hamper it. In any case, many of what are called alternatives have been of long-standing use. One might, alternatively, conceive of a turning away of medical research from the ideas of traditional medicine to those of the 'complementary' medicines; but this is incompatible with the actual expenditure and with the actual activity as reflected in scientific meetings and journals.

(4) Has the earlier success, especially with infectious diseases, simply left medical research with the more demanding tasks of dealing with more difficult chronic, degenerative, functional, and genetic disorders? One need only go back in history to find that it has always felt like this: that the 'easy ' problems have been solved and the remaining ones are intractable; and Bacon's aphorism 110 is yet again worth recalling. It is probably as true now as it was then.

There is, indeed, something rather surprising about these

results, on more general grounds. In the last 40 years, there have been quite remarkable scientific advances. Chromatography, spectrofluorimetry, radio-isotopes, mass spectrometry, electronic instrumentation, computing aids and controls, as well as the advent of plastics and other materials, have both enormously facilitated research and opened up hitherto unavailable methods. Equally remarkable has been the growth in molecular biology and genetic technology, bringing an amazing capacity to identify, isolate, characterize, and 'manufacture' key biological molecules: protein hormones, receptors, ion-channels, antibodies, growth factors, cytokines, promoting and suppressor genes. Many papers in the field claim advance towards a new and more rational therapeutics. It is thus really astonishing that the proportional rate of acquisition of proved new remedies is now less than that of the 1930s.

Some may find the matter of little interest. Some, indeed, may welcome the idea of lower future profits for drug firms. But the reader is reminded again how easy it is for current disease to come to be taken for granted or to be seen as unconquerable (cf. quote from Osler on p. 94), and how improbable new discovery seems before it has been made. It is easy to accept a status quo; but Figure 1 (p. 60) reminds us of what the loss would have been with such a failure of innovation in the past.

## The alternative of human experiment

In earlier sections we have already discussed reasons for distinguishing man from animals, which led us to consider some of the ethics of human experiment, in particular the suggestion that it would be morally better to use a human who is immature or defective in some way rather than some higher animal. Another and older suggestion, sometimes with a vindictive flavour to it, is that it would be better for the experiments to be done on the experimenters themselves. Such suggestions give the impression that experiments are never done on humans and that the idea of human experimentation is a novel one. In fact, it has a long history, although that history has never been adequately written.[16] Part of the story may be found, for instance, in Pappworth's book *Human Guinea Pigs*, a trenchant attack on what he believes to be a largely unjustified use of patients or other humans for experimental work in clinical medicine.[17] His general argument

is that the doctor's first responsibility is the individual care of
the patient who has entrusted himself to him, and that it is
not justified to put such an individual at risk for the general
good without very stringent precautions and thoroughly informed
consent. Experiments should in no circumstances be made on the
mentally sick, the aged, or the dying, or on prison inmates or the
'incurable'. He takes for granted the principle that there will be
full testing on animals before any human experiment is done, and
notes that nearly every code for human experiment recommends
this: e.g. the Declaration of Helsinki (see Appendix 5). He also
offers some explanation of why animal experiment is sometimes
bypassed: namely, that, by law, animals must be anaesthetized for
all but minor procedures, and that human experiments need less
form-filling and record-keeping and are not made the subject of
an annual return to the Home Office. He accepts, however, that
some exception needs to be made for research into mental disease,
where the analogous condition does not exist in animals. The
growth of ethical committees governing clinical experimentation
has gone some way to meet such criticism; but the questions of
patient's consent and of what is justifiable remain actively debated.
In that debate, many of the same criticisms that are levelled at
animal experiment reappear: that modern medicine is losing a
holistic approach, is too 'scientific' and 'analytic', neglects preventive medicine, has been much less successful than it claims to be
or is now actually harmful, corrupts its practitioners, and neglects
patients' 'rights'. The attack on animal experiment and that on
human experiment thus converge in a general attack on winning
new knowledge at all, for no one has discovered a better method
of doing this than by the 'way of experiment'. In fact, human
experiment is like animal experiment—it has its place to improve
understanding and to win practical benefit, and the question is of
achieving the right balance. But that issue is beyond our present
purpose. It is, however, worth surveying something of the pattern
of human experiment, partly to indicate its history and its scale,
partly to bring out the areas where it can be seen to be especially
useful, not indeed as an 'alternative' but as a complement.

Human experiment goes back, of course, to the dawn of history.
The vast bulk of a man's knowledge of traditional remedies came
from trying out on himself or on others the materials in his
surroundings. But so long as there was little knowledge of how the

body worked, only the most elementary observations could be made. The doctrine of the humours, with its categories of dry and moist; hot and cold; earth, air, fire, and water; yellow and black bile, blood, and phlegm reads strangely now.[18] Yet it must have seemed very cogent in its day; for all that could be observed would be largely changes in the 'humours': diarrhoea and changes in colour of stools, change in flow and colour of urine, vomiting, sweating, the production of 'phlegm' from lungs and nasopharynx, and bleeding, together with feeling hot or cold, the relief of pain, and gross behavioural change. But once knowledge of the bodily processes, both in normal function and in disease, began to grow, such observations became recognized increasingly as only superficial signs of deeper mechanisms.

## Some pioneers

The trial of drugs in man continues to this day. Sometimes the chemist or pharmacologist is simply curious to experience what the substance concerned would do to himself, or he may feel that if others may receive it, he should receive it first. Serturner, who first isolated the alkaloid morphine from opium, gives a graphic description of its effect on himself and some students. The Czechoslovak Purkinje nearly killed himself with camphor, and gives a fascinating account of the visual effects of a high dose of digitalis. Christison, in Edinburgh, pioneered the study of eserine, the active principle of the Calabar bean used as an African ordeal poison to test for innocence or guilt.[19] Christison took too large a dose, realized his life was threatened, but was just able to save himself by drinking his shaving water to produce vomiting.

His work proved very important, and led to the substances known as anticholinesterases, from which came a treatment for eye disease, the first means of relieving myasthenia gravis, a whole generation of insecticides used in every garden, and (in nature's customary two-edged way) the 'nerve gases'. The 'nerve gases', in turn, led to a good deal of human experiment. One example flowed from the question of the fate of DFP (the first important 'nerve gas', containing fluorine), after absorption into the human body. It was found, by E. D. Adrian and J. H. Gaddum, after self-administration to test its effect on their own eyes, to be broken down to fluoride in the urine. In fact, from this it was actually possible to make enough hydrofluoric acid to etch glass—a procedure revived many years later by the Animal Liberation Front in a different context!

When the arrow-poison curare was introduced into surgery, to facilitate operative manœuvres by relaxing the patient's muscles, a number of anaesthetists and pharmacologists tried it on themselves. One of these experiments, whose subject was S. M. Smith of Salt Lake City, could claim to be among the most courageous experiments ever done.[20] The reason for doing it was that while it was known that curare penetrated only very slowly into the brain, it was also known from animal work that if applied directly to the brain, bypassing the 'blood–brain barrier', it caused convulsions. It might therefore be possible that with large doses, or during a long operation, effects on the brain would appear. The experiment was, therefore, to give as high a dose as possible to a conscious subject, using artificial respiration of course. The curare progressively paralyses the subject, so a signalling system was arranged whereby movement of any muscle meant 'Yes', and not moving it meant 'No': it would thus be possible to continue until there was no response, even to the question 'Can you respond?', either because the last muscle had been paralysed or because unconsciousness had occurred. In the event, the last muscle to go was the outer part of the left eyebrow; speech had gone long since, and a dose of about three times more than sufficient to paralyse breathing had been reached. The essential observation was made after the subject's rather prolonged recovery; this was that he had been conscious throughout and, indeed, recalled things that the observers had failed to record. If there was any effect on the brain, therefore, it did not touch consciousness, memory, or intelligence.

## Pain

This last experiment points to one area where human work offers a unique advantage: namely, that the subject's testimony can be used. Therefore, whenever functions such as sleep, memory, mood, 'drive', cognitive capacity, or relief of pain are involved, human study is obligatory, as it is for research into mental disease. There has been a great deal of human work on pain and sensation. A great neurologist, Henry Head, and a great surgeon, Wilfred Trotter, both became interested in recovery of sensation after nerve injury, and had nerves in their own arms severed, so as to observe the subsequent changes themselves.[21] A physician, Sir Thomas Lewis, a pioneer of what he termed 'clinical science', explored widely the responses of skin to painful stimuli.[22] He mapped out the characteristic 'triple response' that is so familiar with nettle sting (a dusky reddening—'erythema'—at the point of injury, forming a white 'weal' as tissue fluid accumulates and makes it swell, together with a patchy and pinker reddening in the regions around, called the 'flare'), and showed how local injury (such as a hearty slap on the back for sensitive subjects) produced the effect by releasing histamine from cells in the skin. His later book, *Pain*, remains one of the best accounts of its clinical study, and also shows what the physician can learn from a patient's report about it.[23] Another important study was that by C. A. Keele and Desirée Armstrong, who opened up research into what it is, in injured or diseased tissues, that actually generates the pain sensation.[24] They removed the outer layers of the skin by raising a blister on their own or colleagues' arms with a small mustard plaster, and then cut away the top of the blister. This left the blister base, with pain-sensitive nerve endings exposed, on which they tested various chemicals. The subjects were given a bag to squeeze which recorded their sensations on a scale from zero to very severe. They were able to show that a considerable range of natural substances liable to appear in inflammation were potent and rapid in causing pain, graded in intensity with the amount of substance applied. Other techniques used in man have been the 'hot spot', where a patch on the forehead is blackened and light shone on it until pain is elicited, giving a quantitative measure in terms of the number of calories of heat required; or 'ischemic' pain, where the blood flow to the arm is prevented by a blown-up cuff around the biceps, and then work is done with the forearm until pain develops; or with electric currents to fingers or to the stoppings of teeth, and so on. With such methods, new analgesics have been tested, and various aspects of the physiology of pain analysed.[25] Electrical records have been taken from nerves in the wrist as pain is being elicited, and nerves have been stimulated to characterize the properties of the fibres that actually mediate pain. The work has shown that the electrical records are the same in monkey and man.[26] Such work naturally runs alongside clinical study. One potentially fruitful field is among those who suffer some persistent pain, for they can report not only on relief but also on how long it lasts—an essential datum for clinical use. A patient may welcome such an experiment, since his pain is being put to some use.

## Circulation and respiration

A second area of special study has been of the heart and circulation and of the respiration. Man is unusual in his upright posture, and reflex mechanisms have been developed to stop all his blood pooling in his legs and lower abdomen when he stands up after lying down. Momentary fainting on standing up rapidly after a hot bath or the occasional collapse of guardsmen on parade shows the control system being stretched to the limit. These 'postural reflexes' can only be studied in humans (although some investigations have been attempted in the giraffe!). So there is a large body of very fruitful work with procedures that lower blood-pressure (such as blood-letting, suction on the lower part of the body, or emotional stress) or raise it (adrenaline and related drugs or pain), and their effects on blood-flow through various parts of the body. All this knowledge now contributes to the treatment of cardiovascular disease, shock, and hypertension; and our knowledge of the postural reflexes has allowed the development of another technique in surgery, that of the bloodless operative field, obtained by using a suitable posture at operation to drain the blood away from the region being operated on.

The human study of the control of respiration had a different impetus: namely, that the human subject can provide skilled cooperation in complicated experimental procedures. In the days when it was possible to analyse the blood gases only with great difficulty, J. S. Haldane established that by a special respiratory manœuvre one could obtain a sample of air from the deepest part of the lung, where exchange between blood and air breathed takes place, which provided a measure of the amounts of oxygen and carbon dioxide in the blood. He was then able to show that respiration was largely controlled by keeping the $CO_2$ level constant; when the body does work and oxygen is used, $CO_2$ is produced which raises the $CO_2$ level in the blood, and the breathing increases to wash it out. Subsequently this model proved to be too simple; other factors were identified, and their interaction mapped in a detailed mathematical model. Virtually all the work was done by physiologists on themselves.

A similar participation has helped to throw light on the control of limb and eye movement. Here the subject can try to perform various tasks—such as a maximal muscle contraction or tracking a target or responding to a stimulus—during which movement or nerve activity or electric potentials on the globe of the eye or from the skull are recorded.

## Human survival in hostile conditions

A third area, which received particular stimulus in the World Wars, especially the last war, was research to ensure man's safety in the air, under the water, on the surface of the water after shipwreck, or survival in adverse circumstances.[27] For many of the problems, although not all, there was a background of general knowledge of body mechanisms often gained from animal work; but to know precisely how little oxygen, water, or food, or how high a temperature a man could tolerate and still be effective required direct experiment. So the aviation physiologists discovered how high men could fly both without oxygen and with oxygen provided and how to avoid 'aviation bends' when the pressure was very low, and the effects of the sudden bursting of a pressurized cabin. They defined what acceleration the body could tolerate when coming out of a dive or if suddenly ejected or on crashing, and they designed and tested equipment to increase the tolerance. Such work now underpins the safety precautions of a modern airline, the ejector seat, and the lifebelts for survival at sea. Another especially courageous experiment was that of a professor of anaesthetics, E. A. Pask, who wished to verify that a buoyancy suit for survival of a pilot shot down at sea would be sure to keep his face out of the water even if limp and unconscious; so he had himself anaesthetized and given curare to ensure his limpness and then thrown into a swimming bath wearing the equipment concerned.

The diving and naval physiologists for their part discovered man's tolerance to carbon dioxide and high oxygen pressures; they worked out how to minimize the risk of bends with new depths and new underwater tasks, explored how a man could be brought rapidly but safely to the surface from a sunken submarine, and discovered what temperatures he could tolerate and how best to save him from drowning and from death by cold while awaiting rescue. Other research discovered the tolerance to high temperatures and humidity at various rates of work. Related to civilian life were studies in which conscientious objectors also often took part, to discover the minimum amount of food, water, or various vitamins required for health; this produced fundamental nutritional knowledge in the process.

Chemical warfare had been experienced in the First World War, and it was one of the good features of the Second not only that the arsenical war gas lewisite was not used but that an antidote to it was found. Human experience was already available. A new way of neutralizing such compounds was discovered and shown to be effective in animals. The investigators then tested it on their own skins, showing that if applied at once the new remedy (BAL, British Anti-Lewisite, dimercaprol) could virtually abolish the effects; but that with time the damage became irreversible, treatment being almost ineffective after 24 hours. It is probable that some of the pictures of damage to rabbit's eyes, so commonly presented as routine scientists' cruelty, come from wartime work done on the eye in this programme. On the field of battle, it is of course unlikely that a remedy would be immediately available, and it was important to know for how long after poisoning the eye could still benefit. This entailed allowing eyes to get partly damaged, but revealed that benefit could still be obtained up to 24 hours later. That could not have been found out any other way. (It leaves, of course, for individual decision the moral question as to whether it is better to damage a man's or a rabbit's eye to minimize the damage to any creature exposed to this type of irritant in the future.)

## Clinical pharmacology

Today probably the widest field of human experiment is in clinical pharmacology. As techniques have advanced and medical standards risen, greater and greater attention has been directed to adverse effects in patients and to making treatment more precise. This in turn needs thorough knowledge of the amounts of each drug in the body fluids in relation to dose, the speed of its elimination (for if it is persistent, daily doses will lead to steady accumulation), whether it forms by-products with undesirable effects, and how far these vary with age or disease or genetic inheritance. In addition, far more accurate assessment needs to be made of what a drug does in all its actions and how it interacts with other drugs. While mechanisms may have been mapped out in the test tube, tissue culture, or animal, the final establishing of detailed behaviour now needs to be done in the species in which the drug is to be used—man, horse, cow, sheep, dog, cat, or whatever it may be. So far as man is concerned, then, there has been an explosion of experimental work. If one takes a single issue of a monthly journal of clinical pharmacology, one will find scores of experiments in patients and a similar number on normal volunteers. By contrast, in a journal of general experimental pharmacology, around half or more of the papers will be on isolated tissues, extracts, or enzymes, and the others either on wholly anaesthetized animals or a few involving conscious animals in work on some central nervous problem or on animals subjected to some drug treatment and then shortly afterwards anaesthetized and killed and the tissues sampled.

Some readers may find this account of experiments by scientists on themselves and on other humans irrelevant to the issue of animal experiment; they may not care what scientists do to each other so long as they leave animals alone. They might note that only *some* experimenters do experiments on themselves, and ask 'If some do this, why not all?' They might conclude that scientists are just a cruel lot, as callous about man as about animals, and that it should all be stopped. They may deplore human experiment under the stimulus of war or for anything connected with industrial profit. To meet such comments would be to go over old ground. More important, the purpose of outlining the scope of human experiment was rather for the reasons indicated earlier. First it shows that any belief that biomedical scientists as a whole never put themselves, but only animals, at risk is false. Secondly, and more constructively, one can see how human and animal experiment fit together. It is not *only* that man is more valuable than animals, it is also that each species, man included, is particularly appropriate to tackle particular questions: and when experiments can be done only in man, man is used. The pattern of work, in short, is now a dynamic mixture of non-animal, animal, and human experiment, with the choice made according to the investigative need.

## Summary

1. The risk of the unnecessary repetition of experiments and the forces minimizing it are reviewed. The risk was found to be small by the Littlewood Committee, and available methods of data retrieval now available lead to the same conclusion.

2. The long history is surveyed of the development by the animal experimenter of methods using not whole animals but isolated organs, isolated tissues, tissue culture, isolated cells, subcellular components, modelling, and structure–activity relationships.

3. The view that 'alternatives' can entirely replace animal experiment represents an intensely 'reductionist' philosophy, relying on fragments to represent the whole.

4. Evidence from Home Office reports, Medical Research Council reports, and the scientific literature all show the wide extent to which such methods are used.

5. Over the period 1930–90, drug innovation and animal experiment rose steadily in parallel until about 1960, plateaued, and have now declined in parallel. The pattern indicates that drug discovery is now suffering from lack of animal experiment.

6. It is sometimes implied that there is little experiment on man. The extensive range of this in drug discovery and in work on pain, circulation and respiration, personnel research, and clinical pharmacology is reviewed.

7. It is argued that the correct approach for biomedical research should be a rational mixture of non-animal, animal, and human experiment, as appropriate to the investigation concerned.

# 9

# Man and mouse

It is sometimes contended that only work on humans can be relevant to human knowledge or benefit, because of the biological differences between men and animals. If we took this view, all medical research should be done on man and all veterinary research on animals, with neither able to benefit from the other. It seems illogical to claim simultaneously that man is so like animals that to distinguish between the two is like sexism or racism and that man is so unlike animals that research on them is irrelevant to human biology. Much more important, however, is the fact that if the argument in Chapter 2 is accepted, then this logic-chopping is beside the point: man shares much of his biology with animals, but yet has established, by his capacity to accumulate and build on his experience, a qualitative difference. The real issue is that of the respects in which their biology is close and those in which it is divergent.

An interesting study by Searle *et al.* in 1989 lists 322 genetic loci that are homologous in mouse and man.[1] They cover a wide range of biochemical and physiological functions, largely 'housekeeping genes'. Of the loci, 102 are known to be associated with human hereditary diseases, including a form of albinism, Duchenne and Becker muscular dystrophy, a form of colour blindness, and xeroderma pigmentosum A. Although this represents only a sample of what is to be discovered about the two genomes, it illustrates the extent of useful overlap. It is considered that the common ancestor of the two species lived about 70 million years ago, with about 350 million generations separating the species today.

One obvious point of divergence is in the area of the highest nervous functions. The observation that man and mouse cells can be fused to form viable hybrids shows that the whole intracellular machinery is compatible at this level. Yet it requires only a glance at a microscope slide or neurological atlas of mouse, rat, dog, monkey, and human brains to see that while elementary mechanisms could be studied in any of them, the more complicated functions can be assessed only with the more highly organized structures. As so often, the two approaches are complementary. Some remarkably primitive systems have been invaluable in mapping out elementary mechanisms, simply by virtue of their relative simplicity and accessibility; examples are groups of nerve cells in the leech or the sea slug, the nerves to crayfish or lobster muscle, the junction between crayfish nerve fibres, and the famous squid axon. From these, various patterns of the way excitation and inhibition are transmitted within a cell or from one cell to another can be identified, and tests for characterizing them found, which can then be applied in more complex situations. But when one comes to the problems of how activity from scores of different sources is integrated, the higher systems must be used.

## The surface-area rule

A second major field of difference arises from what is called 'metabolism', the way an organism obtains energy from its food, deals with foreign substances, eliminates waste products, or responds with changes of heat production to environmental change. There is a general rule that one needs a dose of a drug about fifteen times higher in terms of body weight in a mouse than in a man to produce comparable effects.[2] How then can the two be regarded as in any way comparable? The answer is quite a simple one and rather interesting generally, although it may not be familiar. It may be remembered from schooldays that the volume of a sphere is $\frac{4}{3} \pi r^3$, and its surface area $4\pi r^2$. It follows that if one doubled the radius of some specimen sphere to obtain a larger one, then its volume (and therefore weight) would increase by $2 \times 2 \times 2 = 8$, while its surface area would increase only by $2 \times 2 = 4$. In other words, as it becomes bigger, *its surface area diminishes proportionately to its volume*. Conversely, smaller spheres will have a proportionately higher surface area. The same principle operates with shapes that are not spherical. This, of course, is why if you want to dissolve some solid substance in water, you will find that it will dissolve much faster if ground into a fine powder than if left as a large lump or crystal; the weight of the material is the same, but the surface area, at which the solution into water takes place, is enormously increased.

Turn now to comparing a mouse with ourselves—both warm-blooded animals, but the mouse having about 15 times greater surface area in relation to its weight. Because we are warm-blooded, we are continually losing heat from our bodies through the body surface (by radiation, convection, and evaporation of water). Consequently the mouse is losing it proportionately much faster than we are, and so needs more food in relation to weight than we do, and must burn it faster. All its body processes are accelerated, and foreign substances are eliminated faster. This may well contribute to its shorter life-span. A typical result is that any drug which is broken down in the body and requires some time to act is bound to be apparently proportionately less active the smaller the animal it is tested on. But what is happening is revealed if you actually measure the amount present in the blood; there you find that for equal effects the blood levels are similar, despite the higher doses in the smaller animal. It is simply that with the higher turnover in the smaller animal, more must be given to obtain a given level in the blood. One striking example arises if one is examining the way alcohol is dealt with in the body: an alcohol level that could be disposed of in man in around 4–6 hours would take 8–12 hours in a horse,[3] 2–3 hours in a dog, and ½–1 hour in a mouse. As a result, if you try to make mice drunk, it is quite remarkably difficult! You can give them 20 per cent alcohol (about equivalent to port) as their sole source of water, which they then drink during their active period overnight; yet you will fail to detect any alcohol in their blood in the morning. Though man and mouse differ in this way, therefore, it is a difference one understands and can correct for.

This simple point accounts for a fair number of the reputed discrepancies in results between animals and man, for instance in the 'half-lives' of various drugs in various animals—usually mouse, rat, rabbit, cat, dog, monkey, or man. It must be remembered that individuals in any one species always vary in sensitivity; a good working rule in man is that the range of doses of a medicine to be effective in most of the population is about three-fold. Certainly the life-times of many drugs (and hence potency) differ in these animals; and because a drug will seem less potent if it is rapidly destroyed, drug potencies will vary similarly. But they will be found very often to be correctly ranked by the surface-area rule, especially if normal variation is allowed for.

One exception to the surface-area rule is when the drug is not metabolized, either because its action is very swift or because it is chemically very stable. One example is a very potent curare (which tends both to be stable and to act quickly) from the South American jungle, 'toxiferine 1', possibly the most potent known.[4] Its potency is almost the same in frog, mouse, rabbit, and cat; and one may speculate that it was picked out by the Amazonian hunters because of its regular potency whatever the species hunted.

There are differences in routes of metabolism which are not accounted for by the surface-area rule: it is more likely that these depend on the habitual diet. But even then, it is rather that a particular metabolic route is *preferred*, than that it does not exist in other species; and details of the main biochemical pathways can usually be examined in a variety of animals.

Beyond this, however, there are many individual differences between men and animals: given what was said earlier—that each of the different species stands not in a genetic continuum but at the end of its own evolutionary chain—this is not surprising, particularly if one also considers differences in diet, sexual cycle, and 'life-style'. In fact, these differences are probably helpful in the end rather than the reverse, provided they are understood. Man's biology varies considerably: as a vet once remarked to me, 'Compared to the average laboratory animal, man is an appalling hybrid!' Some examples in man are the special liability to sickle cell anaemia and thalassaemia in the African and Mediterranean area; the low acetaldehyde dehydrogenase of the Japanese, that makes them flush more readily with alcohol; the latent myasthenic; the subject lacking the enzyme non-specific cholinesterase in the blood, making him or her especially sensitive to a particular muscle-paralysing agent; the subjects who are 'slow acetylators', who inactivate certain drugs more slowly than the rest of us. The similar variation among animals can be helpful, for it provides a chance of finding an animal model that is comparable for any particular human idiosyncrasy. An interesting case involves vitamin C, where man, in his need for it in his diet, resembles a few animals (such as the guinea-pig and an Indian bird, the red-vented bulbul), but differs from the great majority. The guinea-pig has thus been of great importance for nutritional studies. But it does not stop there, for when one considers the *function* of vitamin C and its requirement in diseased or stressful states, it is striking to find, in animals like the rat that can make it for themselves, what large amounts they can manufacture; and this has stimulated interest in the use of vitamin C at doses higher than those required just to prevent scurvy in normal adults. It illustrates yet again how biological information does not exist in atomic fragments, all isolated from each other, but forms a fascinating whole.

But one should not overestimate these differences. We have only to look at an account of evolution or at textbooks in comparative anatomy to see how much we have in common: hearts, lungs, kidneys, brains, endocrine glands, nerves, muscles, digestive systems, all built on the same plan. This homology goes back still further as one moves down to the biological elements like the nucleus, the mitochondrion, or the cell membrane, out of which the higher organisms are built. One result is that if one opens the *British Pharmacopoeia* at any point and asks, 'Is the type of action of this drug the same in an animal as in man?', the answer in all my own trials has always been 'Yes'. The only differences that appear are in dose required, duration of action, sometimes in the way the action manifests itself, and sometimes in side-effects. The same conclusion follows if one simply compares the drugs used by the doctor and the vet: the overlap, as explained in Chapter 5, is very great indeed.

## The case of penicillin

One last example may be quoted. It has been said that if penicillin had been tested on a guinea-pig, it would never have been introduced into medicine, because it is so toxic to that species. It has even been said that Florey 'admitted' this. There is no historical basis for this conclusion, and Florey's own words are significant:

If we had used guinea pigs *exclusively* we should have said that penicillin was toxic, and we should probably not have proceeded to try and overcome the difficulties of producing the substance for trial in man, difficulties that seem to me now in retrospect even more fantastic than they did at the time.[5]

The crucial word (my italics) is 'exclusively'. Florey was too good a scientist to restrict himself to a single text—and Fleming too, for that matter. Fleming had already found his penicillin preparation harmless to white cells and no more toxic to rabbits than the growth medium. Florey in turn found his extract harmless in tissue culture and to mice, rats, and rabbits.[6] Nobody would have abandoned a substance with such outstanding antibacterial properties and so innocuous to cells and to three species, even if the impure extract was found harmful to a further species. The toxicity to guinea-pigs was known early on, and there is no evidence that it hampered development in any way. The real difficulties in the early stages were obtaining enough of the material, the painfulness of the injection with the still impure material, and the febrile reactions ('pyrogenic responses') also due to impurities. (One of my own early tasks for the MRC in the 1940s was to test penicillin preparations on rabbits for freedom of pyrogenicity, as required by the regulations of the time.)

But the guinea-pig result may be regarded in a different way. The first major drawback with penicillin arose when some patients died from allergy to it. Today, therefore, penicillin sensitivity is usually investigated before treatment starts. It is interesting, however, that the guinea-pig is like man in two particular ways: in its inability to make its own vitamin C (so that we can both get scurvy) and in its allergic responses and in the way these manifest themselves, particularly in suffering bronchospasm comparable to an asthmatic attack in man. The effect of penicillin in the guinea-pig was not obviously an allergic one, but it could well have provided a warning that allergy to penicillin would appear. Another oddity of the guinea-pig is that its intestinal flora are (rather unusually) very sensitive to penicillin.[7] It is thought, therefore, that its effect in this species provides an analogy to the intestinal problems due to invasion by new organisms that later became familiar in man with broad-range antibiotics. Toxic effects tend to be regarded simply as adverse phenomena of no further significance; in fact, they are often of considerable importance as pointers to unexpected possibilities.

## Summary

Man and mouse and other animals have an immense amount of biology in common. Some differences can be recognized, with obvious causes, such as varying extent of development of the nervous system, different metabolic rates, different diets and dietary needs, and different sexual cycles. When other differences arise, they are themselves of considerable interest, and can open new possibilities. That may sometimes make extrapolation from animals to man or (for veterinary work) from animal to animal more difficult; but they are also part of the seed-corn of future discovery.

# 10

# Toxicity testing

The importance of toxicity testing and its contribution to human and animal well-being have been very seriously underestimated, not merely by particular groups, but by the community as a whole, and we must start by explaining this.

Man has come, over the centuries, to lead an increasingly varied life, which constantly brings with it new opportunities, new activities, and new dangers. This will only stop when he becomes less exploratory and less inventive, of which there is as yet no sign. So safety precautions grow, which we take largely for granted, until disaster strikes. Tests of aircraft safety or radioactive emissions, of fire-proofed fabrics or parachute harness or climbing equipment, of the sterility of surgical equipment or of frozen foods, of the safety of electrical apparatus or gas cookers, of the safety of a caisson, a bridge, or a scaffolding—the list is endless. No test is infallible. Some tests fall to be done by the individual, some by a manufacturer, some by a public authority.

Among these tests are those on the safety of new chemical substances emerging from chemical research. There are two categories: (1) substances that may be used in human or veterinary medicine and (2) non-medical substances, very numerous and mostly strange, used especially in manufacturing, although their products, such as plastic buckets, synthetic fabrics, and dry-cleaning fluids, are very familiar. To these the public may be exposed, either intentionally (e.g. a new paint or stripping fluid) or by accident (e.g. in an accident to a lorry transporting chemicals or in a fire) or occupationally (e.g. in a factory). If a harmful effect emerges, the substance receives the name 'toxic'. The word is rather like 'weed' in gardening, which has been defined as a 'plant in the wrong place'. So, too, a toxic action is one that is not wanted at that particular time. Many toxins have subsequently come to be therapeutic agents, such as curare, eserine, atropine, picrotoxin, botulinum toxin, or nitrogen mustard; others have provided invaluable research tools.

Toxicity testing is particularly difficult, for two reasons:

(1) The public demands absolute safety: that is, no toxicity at all. To provide such an absolute assurance (even if a substance that could lay claim to it could be found) is impossible, because one cannot provide for an infinity of possible accidental circumstances; but that hardly diminishes the demand.

(2) It must virtually always be assumed that exposure of any part of the body to the substance may take place. If something is eaten by mistake, of course, we expect it to be absorbed. But the other surfaces between the body and the outside world (skin, lung, lining of the bladder and the vagina) are all permeable, and can also absorb. Thus, for a wide range of substances, even local application can lead to uptake and distribution in the blood to the rest of the body. This is exploited medically in the transdermal administration of drugs such as nitroglycerine for angina pectoris.

The resulting task for the toxicologist may be put in terms of modern genetics. Our genetic library of DNA codes for something like 50,000 genes, any of which may be 'expressed'—that is, actually be involved at a particular moment with the formation and function of some bodily constituent. Of these genes, some thousands have been identified so far, many of them shared with other animals. A good many are what are called 'house-keeping' genes, whose products deal with the machinery and materials common to all cells. To detect a toxic effect on these, a test with almost any cell would provide general guidance. But many of the genes are expressed only in particular cell types or cell constituents (e.g. insulin in the pancreas), some of which we can still hardly guess at.

This is the fundamental reason why toxicity testing must involve whole animals. It is also the reason why, in searching for any toxicity, complete and careful examination of the whole animal and its behaviour is needed, and why, for effects on its organs, full pathological post-mortem is required. It is only then that the full range of genes and gene products is available for the showing-up of any toxic effect. Here lies the best chance that any harm can be revealed, before the general public, young and old, become the test-bed. Even then, it must be realized that some genes are normally under the control of other, 'suppressor' genes, and only become expressed under special conditions, such as disease and pregnancy; then unexpected toxicity appears. This necessarily broad approach, along with the potential for revealing still unsuspected mechanisms, is what gives toxicology its attraction and openings for new discovery.

The point was put more colloquially by the late Dr Alfred Spinks, research director of ICI, in evidence to the Home Office Advisory Committee, that 'a compound may affect any of 100 organs, or 1,000 cell types, or 10,000 enzymes, or any interaction of them'.[1]

It will be realized, of course, that toxicity to the products of 'house-keeping' genes should be readily tracked, since many cell types would be affected. If a test substance proved strongly positive, then further study would hardly be necessary unless a therapeutic use suggested itself—for example, against cancer—which required estimates of dosage and side-effects. It is especially with *negative* tests, however, that the problems with cell or tissue culture tests arise.[2] How can they be relied upon to represent the whole organism? Here, indeed, is the crux of toxicity testing. *How far can a negative report from a given test be relied upon to enable a responsible body to give an assurance of safety to a demanding (and also litigious) public?*

With positive tests, high precision in the testing is sometimes obligatory: for instance, with anaesthetics or curare-like substances used to relax muscles during surgery or some cardiac drugs or drugs used in resuscitation—in fact, with any drug which can do serious damage and whose useful dose is close to the dangerous dose. Individuals always vary in sensitivity, so one must also have some idea of the safety margin.

But generally, great precision is not needed. What *is* needed, however, is some knowledge of the nature of the toxicity—that is, what signs to look out for—and some estimate of toxic dose, so that appropriate care is taken in handling or transport.

One final point was made nearly 500 years ago, by Philip Theophrastus Bombast von Hohenheim, otherwise known as Paracelsus. Born in a mining and smelting region of Switzerland, he turned to medicine, and developed a system including the use of metals (e.g. mercury for syphilis) and other chemicals. These were stigmatized as dangerous by his contemporaries. Part of his reply is famous:

If you wish justly to explain each poison, what is there that is not poison? All things are poison, and nothing is without poison: the Dose alone makes a thing not poison. For example, every food and every drink, if taken beyond its Dose is poison.
[The original German of the key phrase is: 'alle ding sind Gift, und nichts ohn Gift; allein die Dosis macht das ein ding kein Gift ist.']³

That effects depend on dose is fundamental in all pharmacology and therapeutics, as well for toxicology; hence the importance of discovering the level of toxic dosage.

## General toxicity and the LD50

It was to meet such needs that the so-called LD50 was developed, and it is important to understand how this came about. Before scientists understood how to deal with the facts of biological variability, attempts were made to estimate such quantities as 'the minimum lethal dose', 'the minimum curative dose', or the 'maximum tolerated dose' in circumstances where evidence about liability to cause death or damage was needed. Examples were digitalis, whose potency varies with the particular crop of foxgloves, and toxins (such as those involved in diphtheria or tetanus) extracted from bacteria used for the preparation of antitoxins. These tests entailed, as the names implied, trying to find doses that *just* managed to kill or cure or *just* failed to kill an animal. The fact of biological variation meant that if the test was conducted on a group of, say, ten, a given dose near the threshold would sometimes kill none, sometimes one or two, occasionally more, leading to much confusion.

Tests of this sort are called 'quantal assays'. The term 'assay' is by analogy with the assaying of gold and silver. In the tests, the measurement involves a 'Yes or No' result, giving the *proportion* of a group that responds in some particular way—hence the term 'quantal'. A general election may be regarded as a 'quantal assay of political potency'.

It required the insight of J. W. Trevan and later of J. H. Gaddum (both pharmacologists) to see that for mathematical reasons the most accurate region to work with in such assays is around 50 per cent, in the range 30–70 per cent.[4] They worked out methods for estimating the potency at the 50 per cent point as a standard, the result providing an estimate of the dose that would produce a 50 per cent lethality. The improved accuracy of the method meant that fewer animals had to die for a given degree of precision. The term does *not* mean that 100 animals must be used or that exactly 50 per cent must die. The 50 refers only to a proportion expressed as a percentage, and LD1/2 could just as well have been used as a way of expressing the proportion concerned.

An important point about the LD50 is that in its very nature it, too, is subject to variation from one test to another. An advantage of the test developed was that it provided an estimate of 'confidence limits'; these should always be quoted with the value itself, although this is rarely done. This approach represented a considerable advance, and became widely used. One use was to combine LD50 with ED50 (the term corresponding to 50 per cent 'effect') into a 'therapeutic ratio', LD50/ED50, whereby the larger the value, the safer the medicine. It came to be regarded as a standard 'datum' and to be called for by regulatory bodies. In the process, investigators and authorities began to lose sight of its original purpose: namely, to provide an accurate value *where accuracy was required*, and it became used where a much cruder estimate of toxicity would have sufficed. It also meant that lethality *per se* tended to become the key sign of toxicity. With highly potent drugs or those used near their lethal level, there is some reason for this. If one is seeking to identify the safest drug in a series of related ones, it may also be desirable. But apart from these situations, information on the *nature* of the toxic effect and the pattern of its earliest manifestation and subsequent development is more valuable. It also became clear that estimates of LD50 for a substance could vary between laboratories, as a result of factors such as details of procedure or strain of animal. Except in rare cases (some curares provide an example), one could no longer think of the LD50 as a standard value comparable to, say, the melting-point of a substance. In addition, those running the special clinical units dealing with poisoning reported that knowledge of a formal LD50 provided little help in the management of cases.

As a result of such experience, as well as the cost of the tests themselves in animals and resources, toxicologists have for many years been exploring other approaches. Some simply seek to reduce the number of animals used. Another method has been not to aim for a 50 per cent point, but to approach gradually from sublethal doses, accepting inaccuracy of estimate. A third method is the 'limit test'; with relatively harmless substances, as soon as a dose is given up to a limit high enough to give the required margin of safety in practice, higher doses are not used, even if death or overt toxicity has not been encountered.

## The fixed-dose procedure

The most significant development, however, has been from an initiative by the British Toxicology Society. Although this young body was much concerned with other problems, such as training, qualifications, a new constitution, and establishing some kind of publication, it had also devoted much discussion to the question of improving toxicity testing. In December 1982, its committee, after discussing more immediate problems, set up a small subcommittee 'to concern itself with the longer term, with emphasis on the scientific aspects of a realistic alternative to the measurement of the LD50'. The essential task was to find the best way to enable responsible authorities to assign a chemical compound to one of four categories: 'very toxic', 'toxic', 'harmful', or 'unclassified'.

A report was presented in February 1983,[5] including a proposal (due to the subcommittee's secretary) that the acute toxicity of chemical compounds should be assessed by their ability to cause serious toxicity, rather than death. This apparently simple proposal in fact constitutes a radical change of approach. The LD50 test used at that time defined death as the end point, sought to find the dose producing it, and then rank compounds according to that dose.

The new procedure, as currently formulated, is shown in Table 11. Now it is fixed dose levels, four in number, that are first defined, and it is the pattern of toxicity resulting that is determined. Thus, a starting dose for a compound is first chosen on the basis of any existing knowledge. Perhaps that dose does nothing; then the next dose up is tested. Or perhaps the first dose causes a death; then the next dose down is tested. If evident toxicity is found on the first run, then the compound can be classified

**Table 11.** The fixed-dose procedure for toxicity testing and classification

| Dose level | Result | Classification |
|---|---|---|
| 5 mg/kg | Mortality | Very toxic |
| | Evident toxicity | Toxic |
| | No mortality or evident toxicity | Retest at 50 mg/kg |
| 50 mg/kg | Mortality | Toxic (or retest at 5 mg/kg if not done previously) |
| | Evident toxicity | Harmful |
| | No mortality or evident toxicity | Retest at 500 mg/kg |
| 500 mg/kg | Mortality | Harmful (or retest at 50 mg/kg if not done previously) |
| | Evident toxicity | Unclassified |
| | No mortality or evident toxicity | Retest at 2,000 mg/kg |
| 2,000 mg/kg | Mortality | Unclassified (or retest at 500 mg/kg if not done previously) |
| | Evident toxicity | Unclassified |
| | No mortality or evident toxicity | Unclassified |

directly. It will be seen that *death is no longer an objective*; fewer animals should be needed; and whereas with the LD50 test, it is necessary to wait to make sure whether an ailing animal will die or not, here it can be killed as soon as serious toxicity appears. The great question was, would it provide the regulatory information needed as well as the LD50 did?

The sequence of events was then as follows; and it shows what is entailed in such an enterprise.

(1) Consultation with senior toxicologists and retrospective comparisons using data from past LD50 tests.

(2) Debate at a Society meeting, in September 1983.

(3) Publication of the approved procedure in March 1984.

(4) Organization of a collaborative study by British industry, supported by the Health and Safety Executive (HSE). This was published in 1987.[6] The principal results were that in a survey

on 41 substances, in comparison with the LD50 test, only half as many animals were needed; the number dying in the test fell from 41 per cent to 15 per cent; and the toxicity manifested was less intense and for a shorter time.

(5) Presentation of these results to an OECD *ad hoc* meeting of experts on acute toxicity from 14 countries.

(6) That discussion led to an international validation study, conducted during 1988–9 by 33 laboratories in the OECD, making comparison with the then prescribed LD50 method. This was published in 1990.[7] Again, the results were most encouraging. The new test gave 88 per cent of the compounds the same rank as the official LD50 test; all the information on nature of toxicity, time of onset, duration, and outcome required for risk assessment was obtained; fewer animals were used; fewer died; and there was less pain and stress.

(7) After the results were presented in September 1989, the test was accepted by the Commission of the European Communities as an *alternative* to the currently approved method, and international discussions began to extend its use.

(8) In September 1991, after some adjustments, at a meeting in Washington, the European and US authorities finally agreed on a fixed-dose procedure, which will allow the LD50 test to be dropped—a widely welcomed result.

The idea for the new test came from the toxicological community; the requisite analysis, debate, and criticism took place within it; the extensive body of scientific work was done by it; and the fighting through of all the international and administrative problems was done by it. It is an attractive case history: a real international advance in animal welfare which began as an idea of a member of a new small society, but a society that had members with both responsibility and the requisite knowledge, and taking 10 years to carry through.

## Cosmetics, toiletries, and the Draize eye test

Among the uses of animals, none has aroused deeper concern than their employment for the safety testing of 'cosmetics'. A private member's Bill was introduced into the House of Lords in 1977 specifically to make it illegal.[8] Statements were made that 'millions' of animals were used for such tests, and the impression was given of large numbers of animals being tortured to allow industrial profit from female vanity. (These statements were totally misleading; it emerged later that there were then only about 20,000–30,000 animal tests per annum, with about 15,000 experiments on human volunteers. Table 13 gives current figures.) The cosmetics industry remained largely silent, and a good many scientists as well as 'laymen' joined in condemnation of such animal use. Yet such condemnation was not universal, and Baroness Phillips's Bill was not carried in the Lords. Some of the reasons for this become clearer if we look a little more closely at the facts.

There is no clear distinction between medicinal and non-medicinal products. One may use a salve to relieve chapped lips or a cold cream for dry itchy skin in old age. Babies may need some relief for soreness. An adolescent may use an antiseptic ointment on acne. Gardeners and housewives apply barrier creams to their hands to stop them becoming rough. Such instances merge imperceptibly into recognizable medical treatments for dermatitis or skin infection. Equally, the use of tinted powders or creams to hide some blemish have their medical parallel in the more elaborate treatment of port-wine stains and other birthmarks. The practical dividing line is whether or not a prescription is needed for the skin preparation. It is unfortunate that cosmetics and toiletries have been given by some such exclusively frivolous associations; both the dermatologist and the social psychologist know better. So we are concerned not just with an eyeshadow or lipstick, but with toothpastes and deodorants and other substances used daily by millions for purposes which may seem minor but which can be very important in everyday life, and involving application in substantial amounts directly to any part of the body.

Because 'cosmetics' refers to so much more than trivial beautification, it is worth reproducing some definitions. The first is that of a cosmetic, by 1976 EEC Directive (76/768/EEC):

A cosmetic product means any substance or preparation intended for placing in contact with the various parts of the human body (epidermis, hair system, nails, lips and external genital organs) or with the teeth and the mucous membranes of the oral cavity with a view exclusively or principally to cleaning them or protecting them in order to keep them in good condition, change their appearance or correct body odours.

Second is the definition in the US Federal Food, Drug, and Cosmetic Act as Amended (1989):

The term 'cosmetic' means (1) articles intended to be rubbed, poured, sprinkled, or sprayed on, introduced into, or otherwise applied to the human body or any part thereof for cleansing, beautifying, promoting attractiveness, or altering appearance, and (2) articles intended for use as a component of any such articles; except that such term shall not include soap.

Finally, a definition of a medicinal product, from the 1965 EEC Directive (65/65/EEC):

Any substance or combination of substances presented for treating or preventing diseases in human beings or animals.

Any substance or combination of substances which may be administered to human beings or animals with a view to making a medical diagnosis or to restoring, correcting or modifying physiological functions in human beings or in animals is likewise considered a medicinal product.

These definitions do not, of course, remove the problem of borderline products; there is an inevitable 'grey area'. The current DHSS attitude is that it is not the product's actual function but its intended use as claimed by the manufacturer in promotional literature that governs whether it be considered as a cosmetic or a medicine. Clearly, not only a 'mistress's eyebrow' is involved, but the whole body surface of man, woman, child, and baby. To make the issue vivid, it is worth consulting the consumer magazine *Which?*, for its discussion of suntan lotions (badly needed by those with skins with low sun-protection factor), insect repellents (classed as 'cosmetics', but virtually essential in some environments, yet capable of serious effects), contact lens solutions (which use disinfectants), and 'cellulite' removers (which involve rubbing plant extracts into the skin).[9] The whole question of herbal preparations, indeed, raises problems, knowing how their actions can range from the inert via the benign to the lethal, and bearing in mind new ways of using them.

The history of cosmetics and toiletries is important. It goes back, of course, to the earliest ages: a wide variety of perfumes, oils, and salves have been used over the centuries. Some pigments were toxic by modern standards: the Egyptians used antimony and lead sulphides for painting their eyelids black and copper-containing minerals for painting them green. The Romans used white lead on their faces or dusted on to their hair, or took red arsenic for a fair complexion. Such substances are used in various parts of the world to this day, but most modern cosmetics have developed from advances in chemistry. Until about the Second World War, the principal agent used for washing was soap. But then the chemistry of long-chain compounds, that has led us to the modern plastic, began to grow. Early discoveries were first the cationic and then the anionic detergents—artificial soap made by adding the appropriate chemical group to a long paraffin chain. With the anionic detergents, materials of unparalleled cleaning power became available, and the cationic detergents proved to be outstanding antiseptics. But they were also found to be capable of irritating the eyeball. In addition, there had previously been several deaths from a depilatory, 15 to 20 injuries from a shampoo, dermatitis and disfigurement from a hair lacquer, and blindness from an eye mascara. It is very instructive to read the US hearings on the subject, both in 1952[10] and two decades later in 1976[11] when the Eagleton Bill, demanding far more stringent tests, was being considered. One study, in 1944, which hoped to show that the use of a cationic detergent would improve the drug treatment of eye disease by improving penetration of drug into the eye, found instead that it could cause severe eye damage. It is not surprising, therefore, that the US Food and Drug Administration commissioned, through the hands of J. H. Draize and his colleagues,[12] the development of rigorous tests of toxicity to skin and mucous membranes. Their purpose was not confined to toiletries, but covered a range of other chemicals to which exposure might occur, through spillage in a factory, for instance. Tests for sensitization were also included.

If a test does not give definite and reliable results, it is not worth having. The Draize eye test met this demand. The eye of the rabbit was chosen, a fairly obvious choice for ease of application and ease of examination of the eye (it is not true that it was chosen because it cannot form tears or has no nasolacrimal duct: it can and has). A defined procedure was worked out of putting 0.1 millilitre of test material into the conjunctival sac. Observations, based on a clearly defined scale of effect on cornea, iris, or conjunctiva, were made at 1, 24, and 48 hours, and at 96 hours if there was any residual effect. The design of the experiments was arranged to be suitable for statistical test; controlled conditions were arranged; and the actual observer was ignorant of the treatment involved, to avoid bias. Since then, practice has been changed according to the substance concerned, and when appropriate, the amount instilled is left there only briefly and then washed out. The obvious point is worth making, that for the industrialist seeking to develop a new toiletry, the objective is to find compounds that fail to give a positive response, rather than the reverse.

It has been suggested that such experiments should be done only under local anaesthesia. There are three reasons for not doing so. Anaesthesia of the eye is itself an interference which would remove the ability to respond to any casual irritant such as dust or a bit of hair, and anaesthetized surfaces are particularly liable to casual damage. Second, anaesthesia would make it impossible to recognize if the substance produced any discomfort or pain; sensory irritation and tissue damage by no means go in parallel. Third, it is an important physiological fact that the nerves in the skin and conjunctiva themselves participate in the responses to many irritants; and if the eye is anaesthetized, the irritant effect is reduced much below that seen in a normal eye, and any harmfulness would be underestimated.

Considerable effort has gone into trying to find an alternative to the Draize eye test. One special problem is that no other part of the body needs to be transparent, as does the cornea, whose opacity results in blindness; no random tissue culture can meet the need here. But it looks as though some progress is possible. One straightforward step has been restricting the concentrations and amount of test material and the number of animals used.

But a more useful development is to use a serial approach.[13] Thus, materials could be tried first on some *in vitro* test for general cell viability. If there is no significant effect, then, although the material may not be wholly non-irritant, it is unlikely to be severely so; so a test *in vivo* on a 'sentinel' animal could next be done. If the *in vitro* test proved positive, then one would move to one of the new test preparations, using an eye taken from a humanely killed animal, for further *in vitro* test. With this, opacity of the cornea or failure of dye exclusion or thickening of the cornea or dye exclusion from its epithelial cells may be looked for. If the response here were low enough, one could again move to the *in vivo* test, knowing that severe irritation was unlikely. If tests were more positive, and it was necessary to know what levels of the substance were irritant, then one could try *in vivo* tests with a stepwise approach, beginning with low concentrations and small volumes of solution. This would provide guidance for any regulatory study required. If irritation is produced, then washout and local anaesthesia are immediately available measures to annul it. As one reviews such work, it looks as though, even if it may never be possible to eliminate animal testing on the eye altogether, the time is close when the information needed will be available without significant suffering.

To end on a note of contemporary realism, it is worth quoting from a report in the *Times* (19 February 1990) on the proposal by the European Community that cosmetic companies should provide an inventory of all cosmetic ingredients and detailed assessments to support their continued use. Strictly implemented, it would entail testing around 6,000 ingredients, including a good many long-used substances. The increase in the amount of testing required is clearly unacceptable, and the original proposals are now the subject of revision and discussion. The original Cosmetics Directive, which the new proposals would amend was in 1976. A grandfather 'clause', allowing chemicals in use at that time to be considered safe, even if there had been no formal tests, is one possibility. But it appears likely that the more recent the advent of an ingredient, the more stringent the evidence required about its safety. If a new ingredient were manufactured in large quantities, then, of course, the usual tests required to safeguard factory, transport, and other workers, would be needed. As one reads of new aids to skin care, this background should be remembered.

The managing director of 'Beauty without Cruelty' is reported as saying: 'We concede that animal testing is the only option with new ingredients, but we do not use them. We stick with ingredients which have been in use since before the 1976 Cosmetics Directive, which this draft proposal would amend.' He said that the more recent a product, the more likely it is to have been tested on animals. 'No company can claim that it uses only ingredients that have never been animal-tested.' But other manufacturers may be able to claim that, at least, no such tests have been done by them or in their name.

## Irritancy and corrosiveness

The historical survey given earlier included problems of irritation and damage to the skin generally, as well as to the eye. It was to deal with this that Draize formulated a second test, based on application of test compounds to skin over a period of time long enough to detect an allergic response, together with tests for general irritancy, not just on intact skin but also after lightly abrading the skin or removing the outermost layers by repeated application and removal of sellotape.[14] The latter would give some approximation to the abrasions of everyday life in men and animals. Another test was of how far the material is absorbed into the body from the skin. It must be remembered that the field here is wider than that of toiletries; we need to know what care needs to be taken in workshops of all kinds, garages, stores, factories, garden sheds, and the like, where humans and animals are exposed to and handle—or mishandle—extremely varied materials.

The original test included applying the test material to abraded skin under an occlusive cover for 24 hours, followed by 72-hour inspection. A determination to avoid missing any possible irritancy led to criticism that it was too rigorous and overestimated the hazard to man. It has now been modified to 4-hour exposure under a semi-occlusive cover, with no need for abrasion.

But there have remained doubts as to how reliably the results predict action on human skin and how far it gives what is needed for an internationally agreed ranking: that is, from 'corrosive', where full thickness destruction of epidermis down to the dermis occurs (e.g. by caustic soda), down to degrees of irritation with or without reddening, oedema, or scab (EEC has 'risk phrases' such as R38, where there is inflammation; the US Environmental Protection Agency has EPA categories I–IV).

Accordingly, a great deal of work has been done.[15] Examples include use of cultured human or animal skin cells ('keratinocytes') or layers of skin studied *in vitro*. Tests may be of whether the cells admit a dye they normally exclude or leak one they normally retain, or release one of the factors known to appear in inflammation or change their electrical conductivity or stop dividing normally. The work is not only to devise alternatives, but also to analyse more deeply what in fact 'irritation' consists of; and it has brought to light a mass of new knowledge of the physiology and pathology of the skin, a tribute to its astonishing versatility as an interface with the outside world. Very sophisticated methods of measuring changes in the skin have been developed, although one investigator has remarked that 'eye and finger are still better than any instrument'.

The work points, as with the eye, not to complete replacement of *in vivo* testing, but to a system of prior testing with *in vitro* methods which can guide procedure and screen out the most irritant substances from need for *in vivo* test.

## Allergy and immunotoxicity

The problem of allergy to medicines and chemicals has long been recognized, with skin rashes if the drug is taken or sensitization and dermatitis on contact. Some general rules were found; for instance, that if a chemical was rather reactive chemically, but did not react too rapidly and formed stable products, it tended to be a strong sensitizer. It was thought these properties allowed the compound to get into the cells concerned and to combine with a cell constituent long enough to bring about the required changes. With the enormous advances in immunology today, a more complex picture has emerged of the formation of a product recognized as 'not-self', followed by its presentation to the rest of the immune system. The best way of trying to predict such behaviour is being actively studied. Studies with cultures of skin cells of various kinds are throwing light on the processes. But the use of animals will remain for final testing for the time being.

A much more recent development has been testing for something like the contrary response, an impairment of immune function. A genetic defect of the immune system has long been known, and is now much better understood. But it is probably the development of tissue grafting (where a safe immunosuppressant is essential) on the one hand, and the appearance of Aids, where the adverse aspect of immunosuppression shows itself only too vividly, on the other, that has stimulated some much needed study. Again, the complex reactions involved mean that whole animal work is required. It remains to be seen whether, for instance multiple cell cultures may allow some useful type of preliminary screen. A recent symposium of the British Toxicological Society (1991)[16] should be consulted for current approaches.

## Teratogenicity

The thalidomide experience made it obligatory to attempt in some way to detect liability of a drug or other substance to interfere with embryonic development. Much work followed, throwing light on many aspects of organogenesis, and many proposals have been made for test systems, ranging from tests on *Hydra* and fruit-flies, through tests on cultures of cells taken at various stages of development, to tests on organ cultures of samples of brain or limb-bud taken from an embryo.

The problem, of course, is that in the development of these structures, a variety of processes take place, one after the other, so that the possible targets for a teratogen are numerous. It was already known, for instance, that exposure to a teratogen at one moment could inhibit arm development, and a short time later, that of the leg; timing is a crucial factor. It was also known that with thalidomide there could be considerable differences in sensitivity, even within a single species (it was not the case that only one or two species showed the effect), some of this probably being linked to metabolic individuality.

Of the scores of tests developed, perhaps that involving culture of cells taken from defined regions of an embryo (midbrain or limb-bud), with subsequent test for progress in differentiation, has been most hopeful to practical use.[17] But again, the picture is that these *in vitro* methods serve essentially as prior screens, allowing the most toxic materials to be identified without animal testing, but not allowing the reliable identification of non-teratogens. It may turn out that their principal contribution for the time being is simply the light they throw on all the mechanisms of differentiation and growth in the embryo.

## Carcinogenesis

The last major type of test to be reviewed in this chapter is that for liability to cause cancer. An important influence here, as well as a general tendency to 'cancer phobia' such as gave rise to the Nixon initiative in the USA, was the passing of the so-called 'Delaney Amendment to the US Federal Food, Drug, and Cosmetic Act in 1958 (see Appendix 6). This provided, in effect, (1) that there was an onus of proof of lack of carcinogenicity in any food additive, and (2) that if the additive concerned, given in any dose over any length of time in any species, produced any cancer, it would be deemed unsafe. It is probable that if there has been any exaggeration of safety testing, it begins here—notably in the giving of very large doses beyond reasonable expectations of exposure. Not all the criticisms are well-based though.[18] Thus, commonly a 'maximum tolerated dose' is administered; but this is not quite what it sounds like. Rather, it is the highest dose that can be given without altering the animal's longevity from causes other than cancer. Similarly, in assessing the relevance of the dose, the 'man–mouse' metabolism ratio is often neglected, although it implies up to 15 times higher doses per unit body weight in the rodent. But the tests are prolonged and expensive both in animal life and resources, and again there has been much work on other approaches, many of them, of course, directly tackling the intrinsically important questions of mechanism.

One important result has been a distinction between those carcinogenic mechanisms which come from an attack directly on the DNA of a cell, leading to mutations, and those involving an attack on other processes in the cell which may lead to its pathological proliferation, and may be viewed perhaps as a response to biochemical injury. Another important discovery has been that of 'suppressor' genes: that is, genes capable of suppressing the activity of other genes. A carcinogen might thus exert its effects not by direct stimulation of cell growth, but by releasing an over-activity. In neurology, this is described as 'disinhibition'; e.g. strychnine causes convulsions in the brain by preventing the operation of central inhibition. (The story of how the gene product of the p53 suppressor gene was initially thought to be a tumour antigen and then an 'oncogene'—a cancer-producer—shows the complexity of the work.[19] There is now a respectable collection of cases in which a human cancer is associated clinically with the demonstrable absence of a suppressor gene.

Another major discovery was that by Ames and his colleagues that it was possible to detect the liability to produce mutations by using bacteria, particularly the well-known *Salmonella* organism. (Mutants of the organism exist which will not grow on certain nutrient media, but mutation may allow them to revert to normal; the formation of 'revertant' colonies on incubation with some substance is thus an index of the substance's mutagenicity.) Since it may be a metabolite of the substance that is active, it is usual to include appropriate metabolizing enzymes in the incubation mixture. This is, in principle, an admirable type of 'alternative': reasonably cheap, quantitative, quick, and rational as a test for DNA reactivity—even if 'bacteria aren't mammals'.

A promising procedure is thus evolving along the following lines, allowing classification of a substance as carcinogenic (genotoxic), carcinogenic (non-genotoxic), or non-carcinogenic.[20]

(a) Review the chemical reactivity of the substance, and then apply the Ames test. If negative, try other short-term tests of genotoxicity *in vivo* (usually looking for chromosomal changes). If negative again, then test for the non-genetic possibilities. These hinge on some general disturbance of particular cells (e.g. liver, kidney, thyroid, marrow, testis, bladder, skin) that is detectable chemically or microscopically and may result in uncontrolled proliferation. If negative on all these tests, the substance could be classed as non-carcinogenic without prolonged *in vivo* trial. If positive, some statement is also available about type of tumour to be expected.

(b) On the other hand, if the substance is found to be Ames-positive, then confirmation of genotoxicity would be sought, followed by *in vivo* test in male rats and female mice (an interesting outcome of the work has been that this combination gives most of the information on effect of species and sex available from tests on both sexes throughout).

To anyone familiar with regulatory work of any kind, it will be clear that throughout, judgement will be required. But these new approaches combine in a valuable way insight into mechanisms and a minimizing of animal use.

## Is the risk of toxicity enough to justify animal testing?

It has been argued, particularly with regard to cosmetics, that the risk of poisoning is trivial, and that if there is any, it should be borne by the user. But to assess this view, consider some figures from a report in 1976 by the Department of Prices and Consumer Protection, on 'Child Poisoning from Household Products'.[21] It was known that in England and Wales there were 10,500 hospital admissions of children of ages 0–4 for suspected poisoning, and the report estimated a real incidence (allowing for outpatient and general practitioner treatment) of 40,000 a year. The survey was conducted to study in more detail the ingestion of household products, by following up statistics from six chosen hospitals. Of 1,723 cases, 567 (34 per cent) involved household products, of which 554 were traced. Around 100 different substances were identified, of which turpentine (67 cases) and bleach (50 cases) were the commonest. Among toiletries and cosmetics were perfume (26), nail polish-remover (18), aftershave (9), toilet deodorant (8), soap products (5), hair shampoo (4), face cream (4), nail polish (2), and hair lacquer, face freshener, denture cleanser, hairsetting lotion, and false nail glue solvent (1 each). The principal place of consumption was, as one would expect, the bedroom.

It was also reported that one child aged 2 drank a whole bottle of sherry, and that another aged 22 months swallowed an estimated 10 grams of gold chloride powder. Thus, somewhere between 3 and 6 per cent of all such hospital admissions involved just toiletries and cosmetics (according to what is included), giving a national total of 300–600 hospital admissions a year, and a larger real incidence.

Cosmetics and toiletries, of their very nature, are generally of very low toxicity, and, as already mentioned, a 'limit test' for toxicity should suffice. What would be a safe limit? The survey just discussed showed that 80 per cent of the cases involved children between 1 and 3 years old: taking a weight of 12 kilograms as an average, then a child would have to consume about 24 grams or millilitres of one of these substances to reach the usually recommended limit dose of 2,000 gm/kilogram weight—that is, around 2 ounces. It may be felt that this is not so impossible; and indeed, that the size of the safety factor remains a matter of judgement.

Another invaluable source is the National Poisons Information Service, set up in 1963 at Guy's Hospital to provide a 24-hour telephone answering service to doctors, health workers, and emergency services.[22] Over the years it has been able to bring together an unrivalled collection of clinical and toxicological information, often augmented by follow-up of cases. (This showed, incidentally, that household bleach is not seriously toxic, as had been feared.) The enquiry rate is currently over 80,000 per year. Analysis of the 1987 data shows that about 90 per cent were emergencies, and half involved medicinal products. Of the others, chemicals in the home accounted for 30 per cent, industrial and agrochemicals for nearly 20 per cent. In the category of household substances, about 10,000 of the enquiries about non-medicinal agents involved children aged 0–4 years, most of whom had taken the dangerous substance by mouth. Of the non-drug cases, 666 (4 per cent) of household exposures, 295 (16 per cent) of industrial exposures, and 90 (18 per cent) of agrochemical exposures were serious. Behind all this, of course, is the anxiety and distress of the homes.

My first purpose in citing these cases of poisoning, especially in children, is to bring home how far humans are social creatures, and not isolated from each other. One might think that, with cosmetics and the like, the users were the only people ever exposed to them, and that they should be left to the consequences of their own actions. But the facts of social life and accident will always argue against this, just as, whatever the advances in preventive medicine, the surgeon trained to deal with the consequences of strife, natural disasters, and accidents of everyday life will never be out of a job.

Of course, there are those who would say: 'We have enough cosmetics, paints, additives, drugs, foods, drinks, industrial materials—assess what we have and then stop.' This intensely conservative approach appeals only to those, and in those areas, where there is satisfaction with the status quo. It underestimates two things: the deficiencies from which others suffer and the human (and animal) drive to improvement, variety, and choice.

There is a reluctance among some to use animals to protect the frivolous and irresponsible (e.g. the cosmetic user, the drinker, the smoker, the addict). It is a fact, of course, that potentially damaging substances are used on a huge scale, by the foolish and careless, as well as by the observant and prudent; and this use is voluntary. One might expect the prudent to look after themselves; but is it, or is it not, a proper activity of society to seek to protect others from their carelessness or foolishness? Is it right or wrong to provide care for the attempted suicide or the rash motor-cyclist who endangered others in a crash or the alcoholic or the drug addict? If these vulnerable individuals are to be provided for, why not, too, all those exposed to the potential poisons of the household and workplace?

Finally, it could be argued that, better than unreliable animal testing, one should simply do more to collect and analyse human experience as it comes to hand—the equivalent, in the non-medicinal toxicity world, of post-market surveillance in the medicinal field. Certainly, the National Poisons Information Service goes a little way in showing how case histories could be collected. But such an abandonment of the preventive impulse would mean something else: that the test-bed for any adverse effect of innovation would be a random, unsupervised sampling of those who turn out to be vulnerable, especially the very young and the old, identified only by chance enquiry.

The dilemmas presented here should find no glib answers. But there do appear to be signs of one useful approach. If the suffering of an animal can be reduced to trivial levels (say to something no worse than the occasional shampooing of the family pet), many would be satisfied; for it would be merely a part of the general domestication of animals. (That would not satisfy, of course, those who wish to segregate humans and animals entirely from each other.) It will have been noticed how often methods are appearing that allow prior screening of the most damaging materials, and that ways are being found to limit discomfort and damage. This reduction of animal discomfort or damage to trivial levels is becoming an increasingly achievable target.

## How safe is safe?

The purpose of toxicity testing, whether of medicines or of environmental chemicals, is safety for humans and animals. What standard of safety should be sought? The simple question is really a quiverful of other questions, perceptions, and political decisions. Three principal elements are:

(1) A background of known risks, expressed usually as death or illness rate per year per 100,000 (or other convenient number) of population. The risk can be made to seem bigger or smaller by choice of time scale or of population regarded as at risk.

(2) The perception of the risk by the population. Especially important is whether the risk is voluntary or involuntary. Social psychologists have identified 'dread' (e.g. of nerve gas or nuclear energy) and 'ignorance' (e.g. of electromagnetism or DNA tech-

nology) as significant special factors in shaping how dangerous things seem.[23]

(3) The 'value' of a risky activity can be assessed in various ways. Thus the amount of money demanded for work in a risky activity (e.g. mining), or spent in some pursuit (e.g. smoking), or demanded in compensation if damage or fatality occur, or spent by the community on prevention all represent measures of willingness to accept risk.

There is a full literature on the subject, which it is beyond the scope of this book to survey. But certain aspects are worth touching on.

## Starr's analysis

First is a paper by Starr in 1969, which is still worth reading.[24] He worked out two sets of figures for a wide range of activities. The first was the risk, expressed in deaths per man-hour of exposure to it. Thus the risk of hunting (during 1965 in the USA) was calculated from an assumed value of 10 hours for a hunting day, the annual number of hunting fatalities, and the average number of hunting days a year; the result was about $10^{-6}$—that is, one fatality per million man-hours of hunting. With 8,760 hours in the year, that is a little over 1 death per year for a group of 100 hunters.

The figures for automobile (since 1900) and aviation travel (since 1940) brought to light a very interesting point: that the accepted risk fell steadily as the number of people driving or flying increased, and began to level off when about 20 per cent of the population drove and 4 per cent used air-travel. It seemed as though, as an activity became familiar, society began to settle on an approved level of risk with it.

Following on this, it was remarkable that the level concerned was not far from that of the death rate from disease for the whole population, about $10^{-6}$ per man-hour. Also remarkably, the risk for groups hunting, skiing, and smoking are of the same order. Even more remarkable, that fatality rate is only a little below the rate for the forces in the Vietnam War. It was as though the community took, as a general reference point for what was an acceptable risk, that from the general run of disease. (It must be noted, though, that the fatality rate in Vietnam was 10 times higher than normal for the military service age-group, age 20–30; and this no doubt fuelled protest.)

A further observation came from using his other figures, on the 'benefit' from an activity. Looking at wages in relation to risk in the mining industry, it turned out that higher risks of accident were accepted for higher wages, according to a simple 'cubic' law such that an eightfold higher risk was taken for a doubled wage. One would expect a trend of this sort, the acceptance of more risk for more benefit, but 'benefit' is so elusive to measure that this is one of the rare convincing examples.

All the risks mentioned so far have been voluntary ones. A final point from Starr's study was that it appeared that for involuntary exposure to risk (specifically the risk from electric power supply) a 1,000-fold lower risk was required.

An interesting and coherent picture emerges: of the community arriving at an acceptable voluntary risk according to the visibility of that risk, with the risk tending to settle at a level corresponding to natural mortality from disease, and risk being allowed to rise if the benefit rises. But for involuntary risk, only a much lower rate is acceptable. The assumptions in the work have been debated,[25] and the picture as presented in 1969 is certainly oversimplified. But it has not been replaced by any equally coherent scheme.

## The monetary value of a human life

A different approach emerges from discussion of how to allocate resources in a health or public service—another much debated field. One illustration will suffice: a study by Card and Mooney (1977) of the value of a human life as judged by the amount of money that public policy was willing to spend to save it. Three instances were quoted. First, it was estimated in 1968 that if maternal oestriol concentrations were screened in pregnant women, the cost per stillbirth averted would be £50. The method was said not to have been widely used. Second, legislation was introduced to make compulsory the fitting of cabs to farm tractors. This was estimated to save 40 drivers' lives a year, at a cost of £40 for each of 100,000 tractors—that is, £100,000 per life saved. Third, after the collapse of the high-rise flats at Ronan Point, which killed some residents, new building regulations were introduced, implying a value of human life of £29,000,000.

Many such paradoxical examples can be quoted. They do not

help, however, when it comes to making decisions about what levels of risk justify what expenditure on prevention or research. There are frequent complaints about the irrationality of the public; and Starr adduced evidence showing how public agitation could, temporarily at least, distort perceptions. More important than complaint, however, would be a steady effort at education, familiarizing the public with the issues in quantitative terms. The language is similar to that of gambling, which is pretty familiar to a good number!

## Clinical trial

One useful development in the study of the adverse effects of medicines has been the collection of those reported, as a medicine comes into use in general practice. The attempt is made pre-clinically, of course, to eliminate such effects. But nothing can wholly take the place of human experience (or animal experience in the veterinary field). In the past, it was left to the chance of particular doctors feeling it worth while to report some adverse action. Now, however, 'post-marketing surveillance' is seriously undertaken. It is estimated that the clinical trial, before marketing, may pick up adverse reactions with an incidence of 1 per 100 patients.[26] The scheme of monitoring of particular drugs through prescription records in selected practices may manage 1 per 1,000. The 'yellow card' scheme is hoped to pick up incidences of 1 in 10,000. The difficulty is clear; at the latter rate of incidence, many practitioners would never see the adverse reaction.

## Safety factors, detection limits, and risk assessment

A last aspect arises when, for instance, the toxicity of some useful preservative has been characterized, and the question arises of setting permissible levels in food.[27] During the 1940s an informal safety factor of 100 developed; the idea was that man was potentially 10 times more sensitive than animals, and that there was a 10-fold variation in sensitivity among humans. So an 'acceptable daily intake' (ADI) could be obtained by dividing the animal test 'no observed effect level' (NOEL) by 100. This is widely used where there is a threshold for the effect.

But there was the problem of possible carcinogenicity, where there is doubt about there being any threshold—for example, of a growth promoter used with livestock. Ignorance as to a threshold steadily led to more stringent rules, up to a 5,000:1 safety factor, and then to the politically attractive step that there must be *no* carcinogenic substance detectable at all (see the Delaney Amendment). That was, by the techniques of the time, not too unreasonable. But then, initially with the use of radioactively labelled materials and later with other methods, analysis became amazingly sensitive and became able to detect substances at far below biologically relevant amounts. Indeed, we can now be seen to be surrounded by carcinogens everywhere—in minute traces. (Grilled or barbecued foods are particularly terrifying!) So a shift in policy took place, and it was proposed that a residue would be allowed that did not incur, using suitable extrapolation, a lifetime cancer risk of more than 1 per 100 million, later reduced to 1 per million. (The overall deaths in the population from cancer run at around 1 in 500 per year, or around 1 in 10 lifetime risk—one has to die of something.)

This meant a turn back from simply relying on chemical analysis, to needing some quantitative assessment of carcinogenicity. One ends with a now familiar picture: *in vitro* methods that may help to exclude potential carcinogens, but we must rely on animal tests to provide safety assurance. Two meanings of 'safe' have emerged: one-hundredth of the 'no effect dose' and a dose carrying a less than one in a million lifetime cancer risk (about a one hundred thousandth of the natural risk).

# Summary

1. The difficulty of toxicity testing arises from (a) the diversity of substances to which men and animals may be exposed and of circumstances in which exposure may occur; (b) the fact that any exposure to a chemical substance is liable to involve any part of the body, through its absorption and thence general distribution; (c) the fact that assurance of safety can come only after a testing procedure in which a whole organism is exposed since the human genome of 50,000 genes cannot be represented by fragments of it; (d) the public demand for absolute safety.

2. The LD50, an important advance in its day for testing general toxicity, is needed only in special cases. The 'fixed-dose procedure', initiated by the British Toxicology Society, is described. Death is abandoned as an 'end point', and the informa-

tion required can be obtained with much saving of animal life and of severity of effects.

3. The fields of cosmetics and toiletries, irritancy and corrosiveness, allergy and immunotoxicity, teratogenicity, and carcinogenicity are reviewed. In all of them a picture is emerging that, while animal testing is going to be required in the foreseeable future wherever an assurance of safety is required, yet cell culture and other *in vitro* methods are beginning to offer valid methods of preliminary screening of harmful substances. One is justified in hoping that, in time, more serious toxicity can be identified *in vitro*, and animal test be required only for low severity or negative tests.

4. Some of the field surveys of poisoning and adverse reactions are reviewed.

5. The concepts of risk and safety are discussed. It seems that the public arrives at a standard of acceptable risk according to its perception of an activity; this standard tends towards the ordinary risk of mortality from disease (one fatality per million man-hours). Dread, ignorance, and propaganda can influence the assessment. The apparent value of a human life, judged by the amount spent on prevention, can vary nearly a million-fold. The working acceptable exposures in toxicology are (a) one-hundredth of the 'no effect dose', where there is a threshold; (b) the dose which brings a less than one in a million lifetime risk of cancer (the natural lifetime risk is about one in ten).

# 11

# Striking the balance

We come at last to the way in which the balance is to be struck between benefit reaching into the future, on the one hand, and the suffering that may be entailed, on the other. From what has already been said, we could focus on three requirements:

(a) that if experiments are to be done, they must be 'good science', well designed, tackling significant questions, shaped so as to produce definite answers; for if science is not good, animals are wasted;

(b) that the possibility of suffering is to be taken into account, and minimized;

(c) that the balance is to be struck responsibly, with some means of assurance for the public that this is so.

In this chapter, therefore, some account will be given of what happens in practice as regards the experimental procedures and the numbers of animals used. This will refer to the situation at the time of writing, although practice is continually changing.

## Good science

Science is essentially a social activity: its purpose is not merely the acquiring of scientific knowledge, but also its communication. Some investigators welcome early discussion and criticism of work in progress; others prefer to wait until a fairly full 'story' can be told. All have depended on the communication of their work by earlier or contemporary investigators. If the work is neither discussed while in progress nor reported when complete, it is as though it had not been done—except perhaps for some insight or knowledge gained by the observer, which may assist his own later work. One may think of the scientist as a member of the international community in his own day, as well as of a community reaching back into the past (cf. pp. 21–4) and forward into the future; or some may view him simply as a man intent on his own aggrandizement. Either way, the fruit of his work comes only with its communication.

That communication comes essentially at two points: the first when he is planning and seeking resources for his work; the other when he reports his results. At each point it is subject to intensive criticism of the same general type, whether he is submitting a proposal for a grant, giving a seminar to his colleagues, giving a paper to a learned society, or submitting a paper for publication. It will rapidly be pointed out to him if similar work has already been done or if, from existing knowledge, the outcome is already obvious or trivial. Any defects in technique, in assumptions, or in logic are seized on. The fact that others are already engaged in similar work (which, if not yet reported even in a preliminary form, may not be known to him) is commonly indicated by grant-giving bodies and can lead to a useful collaboration. If the question asked is scientifically insignificant compared to the resources sought, that is a quite common reason for rejection of a grant application. Equally, rejection is common when a good question is asked, but the experiments planned can be seen to be unlikely to lead to a decisive answer. Both in the giving of grants and in the adjudication of papers for publication, the method is that of so-called peer review, whereby the opinion is sought of those familiar with the scientific issues and the technical difficulties.

Judgement of this sort can be made only by other scientists. Laymen may look at the titles of published papers, and find them strange, comic, or apparently trivial. Such titles may even draw the attention of legislators, and be scoffed at in the cause of reducing scientific expenditure. What effect this has on the custodians of national finances is not clear; but it is obvious that it is no more reasonable to judge a paper merely by its title and without the scientific knowledge required than to judge a piece of legislation by *its* title and without any knowledge of the situation for which it was framed. Some cases of the misunderstandings that can arise are outlined elsewhere (pp. 215–9, 259). The difficulty in correct appreciation of a scientific project is not, however, restricted to laymen. It occurs also among scientists, some of whom have distinct views about the special use of particular approaches and the worthlessness of others. For instance, there is a long tradition of misunderstanding between the clinical and the pre-clinical worker. The former may find it hard to see the relevance of experiments, ostensibly related to medicine, conducted on frog muscle or crustacean nerve, and may find it hard to see any medical relevance, whereas the latter may see the experiments of the clinician as superficial in character and falling short of the rigour he demands. Both are usually wrong, and it is one of the merits of the blurring of the clinical boundary that such misunderstandings are now less common.

One area where experiments with a similar objective, and hence some apparent redundancy, may occur is when some exciting discovery opens a new field—often a technical discovery such as the use of radio-isotopes and the scintillation counter, or of a new way to record nervous activity, or of a new chemical technique allowing far more sensitive analyses of important hormones or neurotransmitters. Then a 'race' may start. But it is a general rule that if you put two scientists at the same starting point, they rarely follow the same experimental paths. Nor would one expect them to, since they differ in the training, reading, and past experience that they bring to bear. Consequently the outcome is often a joint or nearly joint discovery, but reached in different ways and revealing different aspects. The outcome in fact is a sort of simultaneous confirmation, of the type that is in any case essential for an important finding.

It cannot be pretended that scientific investigation, any more than any other human activity, could not be done better. But those who read the scientific journals or who have attended scientific meetings will probably agree that the critique offered to the experimenter on scientific grounds is as severe as any that is encountered elsewhere in life. The giving of your first scientific paper to a learned society is in the nature of a 'rite of passage' or a tribal ordeal; and it will be found that seniority does not diminish the awareness of that critical scrutiny. One can fairly claim that the forces maintaining the scientific standard of the animal experimenter are as powerful as those maintaining standards in any other walk of life. There are few other areas in which such continuous, open, and explicit criticism of one's work is provided.

## The pattern of animal use

*Experimental numbers*

At this point we should turn to the figures provided by the Home Office Annual Report for 1990, on 'Statistics of Scientific Procedures on Living Animals'. It is a complex document, which deserves comment. It contains 29 tables, helped by five figures. Table 2 of the Report, classifying types of work by species, provides over 400 items of information. Table 3 is, in its own way, still more remarkable; less than one tenth of the total of procedures are classified in a $9 \times 30 = 270$-box table, in which over 200 of the entries are blank. The escalation of data-gathering, begun by the Littlewood Committee in 1977 and accelerated by the 1986 Act, on a background of diminishing animal use, must now represent the most elaborate analysis there is of a limited field. It is to be doubted, however, whether the time and money involved for scientists, officials, and printers is commensurate with the information gained, in relation to other demands for expenditure.

Turning first to the general trend of the figures, this was shown in Figure 19. It must be remembered that in 1987, the term 'experiments' was abandoned for 'procedures' which include animals used for production, so that the figures from that date are about 500,000–700,000 larger than those given earlier for 'experiments', due to the addition of animals used for production. Use levelled off between 1971 (the highest point) and 1976, and has then steadily declined, so that in 1990 it was, on a comparable basis, about half the peak value.

Table 12 lists the number of procedures done according to the species used, arranged not zoologically but by number of procedures, together with the percentage of the total. The first point this shows is that, generally speaking, the 'higher' the animal, the fewer the procedures carried out. The great majority of procedures are done on mice and rats. Fish come relatively high because of their use in work on environmental pollution and on infection and immunology. The frog is something of an exception to the customary 'evolutionary' ordering, being used in limited numbers for the obvious reason that it is cold-blooded and differs rather substantially from the mammal. Cats, dogs, primates, and ungulates are each below 0.4 per cent of the total. 'Ungulates'

**Table 12.** Number of procedures performed in Great Britain classified by species, 1990

| Species | Number | Percentage of total |
| --- | --- | --- |
| Mouse | 1,636,332 | 51 |
| Rat | 891,536 | 28 |
| Bird | 247,609 | 8 |
| Guinea-pig and other rodent | 162,459 | 5 |
| Fish | 107,989 | 3 |
| Rabbit | 89,845 | 3 |
| Ungulate | 34,759 | 1 |
| Reptile/amphibian | 13,123 | 0.4 |
| Dog | 11,433 | 0.4 |
| Primates | 5,284 | 0.2 |
| Cat | 4,392 | 0.1 |
| Ferret and other carnivores | 3,612 | 0.1 |
| Other mammal | 821 | 0.03 |
| Total | 3,207,094 | |

(hoofed animals), including cattle, are involved primarily in veterinary study of normal and abnormal body structure and function and for the development of veterinary products and appliances.

The general pattern of animal use will probably be found reasonable, but the question will still be asked: 'Why so many as 3¼ million?' The answer can only be that that is the number that has been found necessary.

There has been some criticism of the use of the word 'necessary'; and it has been pointed out that no logical, analytic, or strict causal necessity has been shown. Of course not. The word 'needed' could have been used just as well, as a dictionary will confirm. The meaning is the same as in an editorial passage in the journal making the criticism, just two pages earlier: 'In most years, our income just about covers our expenditure, but substantial new income is needed if we are to avoid a substantial deficit in 1989–90.'[1] The judgement by a conscientious scientist as to how many animals to use in given circumstances in work to advance medical and veterinary knowledge is of the same kind as the judgement by a conscientious administrator as to how much public money should be spent, and on what, in a particular approach to animal welfare. Neither bureaucratic elaborations against abuse nor questioning of motives are appropriate in either case.

More seriously, the forces working against the unnecessary use of animals and for the increasing use of alternatives, have already been reviewed, as have some of the benefits that have been won in the past. There is no basis for deciding on a priori grounds what is too large or what is too small. The current figures represent roughly one mouse per head of population every 15 years. Is that big or small in relation to the hundreds of millions of animals used for food each year? Or the hundreds of thousands of cats and dogs killed annually as unwanted? Or indeed, the 100 million smaller animals killed by our cats each year? It is not quite sufficient to reply that 'Two wrongs do not make a right'. It is true that if something is wrong, it is not made any better by something else being worse; but most people would assign more priority to dealing with the larger wrong. The last few years, however, have brought evidence of a different kind: that reduction in animal experiment has now reduced pharmacopoeial innovation to the level of the 1930s, as we saw in Chapter 8.

In considering numbers of experiments, comparison with the number of observations needed in areas of enquiry more familiar to the layman may also help, such as opinion or consumer surveys. Few would attach much weight to a survey of, say, 100 people for their opinion on union activity, nuclear waste, or the siting of a new road, as representing a true measure of the opinion of the country at large. Experimental work, because of the variability of living organisms, equally apparent in human political views and in simpler biological responses, is entirely analogous,; and to establish some fact, so that it can be the *basis for future action* is just as demanding of numbers. Indeed, if we review the numbers used in estimating a frequency of opinion, developing a soil for a plant, testing an insecticide, analysing the composition of paper in old books, or finding out the potency of an anti-diabetic drug, they are all of the same order—sometimes smaller, sometimes larger, according to the precision required.

Figure 20 (from the 1990 report) gives an overview of some broad areas. Toxicity testing has been discussed in the preceding chapter. Table 13 shows the number of procedures analysed, with the numbers under various categories of interest summarized.

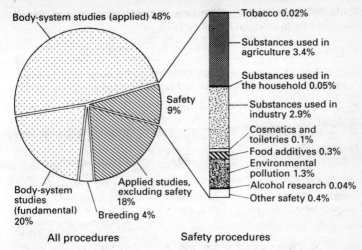

**Fig. 20.** Number of procedures by primary purpose, 1990
From Home Office (1990).

It will be noted that a considerable number of procedures came under the head 'selection of medical, dental, or veterinary products and appliances', together with procedures demanded by law here and abroad. The whole of medical and biological research and production accounted for 91 per cent of the total. Industry carries out 58 per cent of the work, with about one quarter being done in universities and polytechnics. Numbers involving the eye raise a special point; far the greater part arise through the use of the eye for propagation of virus for vaccine preparation. The remaining 6,000–7,000 procedures, to cover the whole of ophthalmological and safety research, seems rather small. Radiation (painless) is used to inactivate an animal's immune system, so that immune cells of defined character can be implanted for immunological research. Thermal injury is another instance where, bearing in mind the commonness of burns and the complexity of the reactions to them, studies seem rather limited. A potentially very important field is likely to be the transgenic animal. This is nothing to do with genetic monsters, but with finding ways to remedy genetic defects and the like. With more and more diseases, genes are being identified that are missing or

**Table 13.** Number of procedures performed in Great Britain/ classified by procedural category 1990

| Procedure | Numbers | Comment |
| --- | --- | --- |
| Substances applied to eye | 109,295 (birds) 116,000 (total) | Propagation of virus for preparing vaccine |
| Ionizing radiation | 58,500 | Immunology |
| Cosmetics and toiletries | 4,365 | |
| Thermal injury | 600 | Anaesthesia + recovery |
| Anaesthesia | | |
| none | 2,205,400 | 69% total (81% rodents, 3% rabbits, 1.6% other mammal) |
| with recovery | 529,800 | 16% total |
| terminal | 472,000 | 15% total (92% rodents, 3% rabbits, 3% other mammal) |
| Universities & Polytechnics | 748,100 | 23.3% total |
| Commercial | 1,866,200 | 58.2% total |
| Other (NHS, government, public bodies, etc.) | 592,861 | 18.5% total |
| Production | 530,000 | |
| Breeding | 140,800 | |
| Transgenic + genetic defect | 140,000 | |
| Surgical technique | 13,500 | |
| Education | 10,400 | |
| Medical and biological research and production | 2,910,510 | 91% total |
| Selection of potential medical, dental, or veterinary products and appliances | 1,079,200 | |
| Medicines Act or overseas equivalent | 444,200 | |
| Health and Safety at Work, Agriculture (Poisonous Substances) Act 1952, Food and Drugs Act 1955, or overseas equivalents | 242,590 | |

*Source*: Home Office Report 1990

altered; and the possibility is opening up of replacing the defective gene. But a lot of work is needed in characterizing such genes, studying the conditions of their expression, finding out how safely to transmit them, and verifying that there will be no adverse effects. Some more needs saying about anaesthesia and animals in education.

## Experiments wholly under anaesthesia

These experiments, 15 per cent of the total procedures, must be regarded as involving no more pain or suffering than is involved in the putting down of a domestic pet by the use of an anaesthetic, since the experiments are done wholly under anaesthesia and the animal is then killed. The proportion of such experiments for the main species used is shown.

This category of experiment consists largely of analytic experiments in physiology and pharmacology, and the experimental procedures would be wholly unacceptable if the animals were not anaesthetized. The cat and dog have been important partly for the analogies between their physiology and that of man (particularly as regards heart and circulation, respiratory function, and neuromuscular function) and partly because their size allows experiments that would be impossible in a smaller animal—although as technology improves, work on the anaesthetized rat is becoming increasingly productive. With animals larger than cats or dogs—such as primates, ungulates, and equidae—the cost alone makes it necessary to find a way that yields more information than an acute experiment can do.

A recent case has raised one important matter: the verification of a sufficient depth of anaesthesia.[2] With an anaesthetic such as a barbiturate, the depth of anaesthesia increases for a short time after induction and then steadily lightens, so that supplementary doses are needed if an experiment lasts for some time. Judging the correct depth of anaesthesia is not a simple matter. It might be thought that it was simply a question of abolishing all responses to sensory stimulation; but such a state both constitutes very deep if not dangerous anaesthesia, and is more than is necessary. The history of the introduction of curare in human surgery illustrates the underlying situation. Anaesthetics first dull and then remove consciousness, while both spontaneous movement (e.g. breathing and eye movements) and responses to stimuli are still present.

Even with deeper anaesthesia, the response to the stimuli of abdominal surgery used to be a tightening of the muscles making the surgeon's exploration of the abdomen very difficult. Thus, for operations like that on the gall-bladder, requiring access deep under the lower ribs, it became necessary to give a lot more anaesthetic with corresponding dangers and after-effects. Curare allowed the muscles to be relaxed without having to give more anaesthetic, and the benefit in terms of access for the surgeon and of patient well-being amounted to a surgical revolution. The need for curare arose simply because of the degree of muscle response to stimuli even when anaesthetized. Movement when anaesthetized *may* signal, but does not *prove*, return of consciousness.

It will be seen that, both for anaesthetist and for experimenter, the giving of an anaesthetic requires skill, alertness, and specific training. For the layman, the successive stages of drunkenness provide some analogy—the fighting drunk can be unaware of his actions. But for practice there is no substitute for trained experience of the sequence of changes as anaesthesia deepens or passes off and the way the details of that pattern vary with the individual agents.

## Experiments without anaesthesia

These are experiments in which, if there is any operative procedure, it is no more severe than a simple inoculation or a superficial venesection, corresponding to a polio 'jab' or a blood donation in a human. For the majority of experiments in this category, the induction of anaesthesia and its after-effects would be more distressing than its omission. But there are some experiments that involve some discomfort. For instance, a new analgesic may be tested in various ways. One is to restrain a rat after injection of the drug or control by a suitable route and to lay its tail across a wire, which is then heated by passing electric current through it; the time is measured before the rat flicks its tail away. A second method is to place a mouse or a rat on a plate heated to about 60°C (like a hot cup of tea), and measure the time before it jumps off. The principle with both procedures is that the animal is free to escape when pain is felt. A second category is in psychological experiment, where, for instance, an electrified wire grid may be used in training an animal in some task. The stimulus, however, need only be strong enough to change behaviour, and involves mild pain at most. A human analogy would be the discomfort that leads to choosing a more comfortable chair.

A more serious issue is that of experiments involving 'inescapable pain' of any duration; these have received attention elsewhere in the world. The experiments entailed, for instance, severe stimuli from an electrified grid for a sustained period: the animal stops trying to escape, and there is some evidence that mechanisms are activated in the brain which produce some measure of analgesia. Subsequently, the animal enters a state which could be regarded as a 'model' for some types of depression. Pain of this order certainly occurs in man (for example, causalgia and tic douloureux) and presumably also in animals, and severe depression, rather than pain per se, is arguably the cause of the deepest human suffering. A case for such experiments can therefore be made. But permission has not been given for such experiments in Britain. In my view, this is entirely correct; and it is doubtful whether the physiological information sought could not be obtained in other ways.

Under the previous Act, there were what were known as 'pain conditions', designed to prevent severe pain from being experienced by an animal for any length of time. (These have been replaced by extensive Guidelines, issued by the Home Office together with the Act.) These led some to conclude that severe pain is common in experimental work. One cannot speak about other countries, but in the UK it is in fact very rare, partly because scientists are as humane as anyone else, partly because if it occurred, it would destroy the effectiveness of an experiment through the physiological disturbances thereby caused.

In my own experience I have never seen severe pain inflicted on, or experienced by, an experimental animal, but only on human subjects. My own worst experience was acting as a subject for a colleague, and having a searchlight at *HMS Collingwood* focused on a 1 cm circle of my forearm, so as to discover what skin damage followed application of precisely known radiant energy input. The most severe pain I have inflicted on an animal is, I believe, when shampooing the family cocker spaniel; some soap got into its eye, and lachrymation, spasm of orbital muscles, whining, struggling, and eventual escape from the operator ensued. It illustrates the problem of irritancy in toiletries.

One other experience, when I was starting research, remains

vivid. I was asked by my professor, in the days before sharp hypodermic needles, to inject a rabbit subcutaneously with pituitary extract, to see if it would develop antibodies (an experiment far before its time). My first attempt got nowhere; my second bent the needle; sweating and feeling brutal, I straightened the needle and made a last try. The needle went through the skin, and at that very moment (which I would have expected to have been the worst) the rabbit, so far wholly unconcerned, bent its head down to nibble a bit of lettuce on its tray. Animal behaviour is by far the best guide for estimating animal experience.

## Experiments with anaesthesia for part of the experiment

This category consists of those experiments in which first some procedure more severe than inoculation or venesection is done under general anaesthesia; the animal then recovers from the anaesthetic, and the experimental observations are made. They constitute 16 per cent of all procedures, and again largely involve rodents. With these experiments, as with those without anaesthesia, the assurance is required that anaesthesia during the observations would frustrate the purpose of the experiment. This is, of course, obvious for any behavioural or analgesic test. It is also the case for the whole area of safety evaluation, because anaesthetics themselves have significant toxic actions, and because they could mask in a variety of ways the expression of toxic effects.

One type of experiment is to implant a fine tube under anaesthesia so that samples of blood or some other body fluid can be taken, or lymph drained over a period of time, without disturbance. Similarly, a suitable device may be implanted to allow blood-pressure or blood-flow to be monitored or recordings to be made of muscle, nerve, or brain activity in the unrestrained, conscious animal. Another category is where a lesion may be made to study trauma or the way the body's systems of repair operate or to throw light on the function of some nerve pathway or group of nerve cells in the brain. Thus a tendon may be severed, a piece of bone removed, a suture material implanted to test for irritation and readiness to be absorbed, a nerve sectioned, a local injection made of some substance which inactivates some nerve pathway, a known number of tumour cells implanted in cancer research, a graft of some tissue or organ made in developing new replacement techniques, part of the liver or some gland removed. With each of these, the suffering entailed is first that of the induction of anaesthesia, next that of the post-operative period (which is astonishingly brief in animals, compared to humans), and then of any disability that may arise. Commonly, whatever the reaction being studied, it is not allowed to progress far, partly because it is the early stages that are of most interest, partly because if distress is caused, it complicates interpretation severely. This is an area where miniaturization and other advances in technology are extremely promising.

## Animals in teaching: Pasteur and the 'prepared mind'

In December 1854, Louis Pasteur gave an inaugural speech to mark the opening of two faculties, of letters at Douai and of sciences at Lille.[3] In it he set side by side mastery of the tongues of Homer, Cicero, and Pascal and mastery of the 'transformations that matter could undergo'. Speaking particularly of the sciences, he stressed the vital importance of practical work by students, in addition to lectures and demonstrations. He remarked on how, to know a country, one had best travel in it oneself. He discussed how theory and practice interact, and told the old story of how Benjamin Franklin, when asked what use there was in some purely scientific observation, replied 'What use is a newborn baby?'. He recalled Oersted's experiment, in which on completing an electric circuit by joining the ends of a copper wire to a Voltaic pile, Oersted saw a sudden movement of a nearby magnetized needle on its pivot. While the current flowed, it no longer pointed north. This was the birth of the electric telegraph and much else. He commented: 'By chance, you may say; but in the field of observation chance favours only prepared minds.'

It is significant that this famous remark by one of the greatest of medical scientists was made in the context of education. It has been echoed by scientists of all types. The geologist, botanist, zoologist, forester, agriculturist, ethologist all call for field-work. The physical sciences call likewise for 'hands-on' experience of the appropriate kind. The success of 'hands-on' exhibits at science museums provides another example. So too, it is vital for the young medical student and biomedical scientist to have his or her mind prepared by seeing under his own hands the 'transformations that matter can undergo'.

It is hard to credit that anyone should suggest that medical or veterinary students should be deprived of this training; that they should not learn how to give an anaesthetic or how to cut skin or muscle, or be directly familiar with the control of flow of blood in artery and vein and the beating of a heart *in situ*, the movements of the gastro-intestinal tract, and the mechanism and control of respiration. How can it be right that not until they deal with actual patients (human or animal) will they give their first anaesthetic, make their first incision, take their first blood sample, or deal with the first actually bleeding artery? The discovery that clumsy handling damages tissues and knowledge about how to avoid it cannot be taught on a video tape. Yet practical classes are under severe pressure; and their replacement by videos or films of typical experiments or by computer simulations is repeatedly urged. The latest Home Office return of experiments for 1990 shows that 10,382 'procedures' were for 'education'—about one per year per medical student, to prepare him or her for the whole world of practical medicine.

More than practical skill is involved in becoming competent. The most striking fact about biological events is their variability. It is the aspect of biology to which those trained in physical sciences who change research field find it most difficult to adjust. Another name for it is 'individuality', one of life's markers and pleasures. Individuality, in fact, lies behind all the urging that patients be treated 'holistically', with regard to their own particular characteristics. It shows itself at every point, in all the patterns of bodily response, compensation, and adaptation. It is essential for the student to learn about such variability, its range, and how to handle it—one reason, of course, why biostatistics is so important a subject. To rely on film, video, or computer is to replace individuality with a fixed stereotype, 'the right answer', suitable for regurgitation perhaps to a bad examiner, but no preparation for the real world of medicine.

This is not to say that animal experiment for teaching need be extensive or exclusive. Film, video, modelling, can with care all play their part alongside the lecture, the demonstration, and the seminar. Nor should there be anything merely repetitive about the animal experiment involved. Indeed, such work, in the hands of imaginative teachers, may not only fire the future research worker, but be itself a research starting point.

In the past, practical classes, in which the students themselves were the subjects for experiments, provided very effective teaching methods, complementing the other practical work. They provided a useful introduction to several techniques in clinical practice, and sometimes constitute part of a research project. Examples have been trials of local anaesthetics, drugs used to dilate the pupil for retinal inspection, sedatives and stimulants, analgesics, and autonomic drugs. There is some irony in the fact that these experiments, too, have been eroded by local ethical committees. Does this mean that, soon, the only subjects judged suitable for experiment will be the 'human incompetent' inhabitants of our hospitals?

## Who is it that is insensitive?

It is sometimes suggested that experimental scientists are bound to become 'desensitized' to the suffering (if any) that they inflict on their experimental animals. Indeed, the point has been developed at some length.[4] Similar remarks are made about doctors and their professional jokes; and lay persons seeing a surgeon calmly dealing with a surgical emergency or a vet doing some essential operation might well feel that these, too, were unfeeling and cold-blooded in their lack of evident empathy with their patients. The superficial appearance of lack of feeling is not to be confused with the actual exercise of professional skill and efficiency. The animal experimenter likewise must be professional and efficient in his work. My own experience is that animal work makes the investigator *more* sensitive to animal needs, as he replaces anthropomorphism by knowledge about their behaviour and their physiology and gains experience in seemingly small but important matters about how to handle them. One might indeed wonder how it comes about that those who are trying to discover more about the working of the animal body, with a view to removing ignorance and improving human or animal welfare, should be thought to be *less* sensitive to animal needs than those who seek to hinder such work.

But these considerations seem to go for nothing. Little objective evidence of desensitization is presented apart from a handful of personal testimonies of becoming 'desensitized' by scientific work (although if this leads to concern regarding the issue, there is some self-contradiction). What *is* presented is in effect a long sneer at scientific training, portraying the majority of its practitioners as succumbing to inevitable desensitization, peer-group pressure, curiosity, job security, and the search for eminence, before all else. The only rational content is the statement, in effect, that anyone who disagrees must *ipso facto* be either wicked or defective—the invariable claim of the lobbyist. Apart from denial, there can be no answer to all this, nor to the fantasies about visitors from other planets. But as one hears or reads these opinions, an image appears behind the voice: two sad processions. One is of the victims of diseases like polio or distemper, or of African plains littered with corpses from cattle plague, that would have been, had these views been adopted in the past. The other is a less defined procession of hopefuls in the future, the prevention of whose suffering depends on the animal experiment of today. How could anybody be insensitive to these?

## The context of animal experiment

In this chapter we have examined the forces maintaining the scientific standard of animal experiment; and we have looked at the actual decisions taken and the final outcome in terms of animals used, as shown by the Home Office returns. But something should be said about the general system of control.

First, at the risk of some repetition, it is worth pulling together some of the elements of the context in which the animal experimenter works. The impression is sometimes given that the experimenter works in isolation, responsible only to himself, free to do anything he wishes on as many animals as he wishes. Nothing could be further from the truth (it is much more nearly true for the pet-owner).

Before an experiment can be done, a place must be obtained in what used to be called 'registered premises', now 'designated places', that have Home Office approval. Then stipend, equipment, technical assistance, and running expenses are required. All these, whether in universities, industry, or research institutes, are in short supply. There is, therefore, competition for these resources and close scrutiny of qualifications, programmes, and all the costs, both in departments with regard to their own budgets and in grant-giving bodies in relation to their available funds. With the latter, rates of rejection of applications for support may range, according to the economic climate, up to 70 per cent, with 'alpha' ratings going unfunded. The critique offered by grant-giving bodies extends as liberally to senior as to junior investigators.

Secondly, licences must be obtained. Under the 1986 Act, these are of two types: a 'project licence' specifying a programme of work of specified type at a specified place and a 'personal licence' allowing actual application of a regulated procedure to an animal, as part of a programme covered by a project licence. The purpose of the work must be one of those approved in the Act. Details are required of the programme and the animals to be used, with assurances that alternatives have been considered and that no 'higher' an animal is used than is necessary. The Secretary of State, in deciding whether to grant a licence, shall 'weigh the likely adverse effects on the animals concerned against the benefit likely to accrue'. An application must be endorsed by a holder of a personal licence with personal knowledge of the applicant's qualifications, preferably someone in a position of authority. Commonly, the application is discussed with the local Home Office inspector.

Given approval thus far, experimental work may start. It is then subject to visits, without notice, by the local inspector, who also inspects the animal-holding provision and may talk to animal technicians. At the end of each year, a return must be made to the Home Office of animals used in experimental work, the use being categorized under a range of headings (Tables 12 and 13 give some idea of the range).

Other features of the new Act are regular inspection of both experimental and breeding facilities, restriction of animals to those from registered breeders if possible, increased veterinary participation, new rules on re-use of animals, and an independent monitoring committee with power to volunteer advice, now called the Animal Procedures Committee. (In passing, it is to be noted that the advisory committee 'shall have regard to the legitimate requirements of science and industry and to the protection of animals against avoidable suffering and unnecessary use in scientific procedures'. There is nothing about 'prevention or treatment of disease, ill-health, or abnormality in man or animals', as there is in the Secretary of State's terms of reference. The committee is apparently entitled to ignore the needs of medicine.)

That is an outline of the formal procedure. But scientists do not work in isolation. They are continually exposed to the opinions of their colleagues, graduate students, students, and technicians. A primary objective of science is that it is shared with others; and while this may be dismissed as the 'publish at all costs' disease, it is in fact at the heart of scientific work. Accordingly, the work is reported in departmental seminars and to learned societies. For the work to be assessed, the procedures used have to be described. If the procedures are felt to be in any way inhumane, they may be questioned by members of the society. The work is then written up for full publication after peer review. I can only speak of British practice, but here at least, account is taken of the humanity of experiments done, and publication may be refused to reports of those regarded as inhumane. After the work appears, it is then subject to general scrutiny; welfare societies sometimes publish analyses of their findings.

## Is animal experiment adequately supervised?

It has been contended that because there were no prosecutions under the 1876 Act, the inspectors must have been failing to provide the control required, and that in effect experiment had been uncontrolled. Historically the statement about prosecutions is not correct.[5] There have been, so far as I am aware, three prosecutions, which are worth a brief summary. The first was in 1876, when a certain Dr Arbrath was prosecuted for advertising a public lecture on poisons in which (unspecified) experiments would be shown. The advertisement went to press before, and appeared three days after, the passing of the Act. Although in the end no experiment was performed, Dr Arbrath was convicted with a nominal fine. It is to be noted that he belonged to the local branch of the SPCA (as it then was), and that it refused itself to prosecute. A second case was in 1881, when the Victoria Street Society (precursor of the NAVS) prosecuted Dr David Ferrier for performing experiments on the brain while unlicensed and uncertificated for such experiments. The prosecution failed, however, because the operations involved were in fact performed by Dr G. F. Yeo, who held the required licence and certificates. Ferrier is famous for his work on cortical localization, fundamental to modern neurosurgery; and a Royal Society lecture is named after him. A third case was in 1913. This was a prosecution by the RSPCA of Dr Warrington Yorke for cruelty to a donkey. It involved an experiment in which a drug possibly useful against sleeping sickness produced a type of paralysis. The prosecution failed, because Dr Yorke was properly licensed and the suffering involved was not judged to be unnecessary. Dr Yorke, later a Fellow of the Royal Society, was a pioneer in tropical medicine, and with Kinghorn discovered the insect vector of one of the sleeping sickness parasites (*Trypanosoma rhodesiense*), an essential piece of knowledge for its control. One further case is of historical interest, since it probably contributed to the movement which resulted in the 1876 Act. This concerned Dr Magnan, a French investigator who demonstrated at a public meeting of the British Medical Association in 1874 that whereas injection of alcohol into a dog produced anaesthesia, injection of absinthe produced convulsions. He did this to draw attention to what are now well-recognized dangers of absinthe. The RSPCA prosecuted under Martin's Act (the first act in this country dealing with cruelty to animals), but Magnan had by then left the country.

In fact, prosecutions are rather beside the point. There is an immediate sanction available: namely, the withdrawal of licence or certificates from an individual or from premises. For the vast majority of scientists who do animal experiments, such a withdrawal would radically affect their careers and could lead to some losing their jobs. No data exist on how often this has been done, but it is clearly rare. This could be taken as reflecting the lack of abuse of the law; or it could show the ineffectiveness of the inspectorate. The former seems the more probable, in the light of the Annual Reports which record each year their detection of a series of infringements. In the great majority of cases, these are technical, such as allowing a licence or certificate to run out while experiments continue. The essential point to which the inspectors' attention is directed is whether any unnecessary suffering was caused by the infringement. Appropriate warnings are given, and in two recent cases the facts were brought to the attention of the Director of Public Prosecutions. Infringements sometimes arise in the course of work by visitors to this country from other countries with less stringent regulations.

It has also been contended that adequate supervision cannot possibly be conducted by 15–20 inspectors. This implies some misunderstandings, and some figures are quite enlightening, if one compares 1982, say, with 1988. First, experiments take place in a relatively limited number of recognized places: 518 in 1982, 381 in 1988. The inspectorate made 6,531 visits, mainly without notice, in 1982; 7,640 in 1988 (figures are not given for later years). The number of centres has therefore declined by 28 per cent; visitation has increased by 17 per cent; and the average number of visits per year to a centre has risen from 12.6 to 20. In 1982 there were 11,800 active licensees, an average of 23 at each centre. (This information also is no longer given—only project licensees.) The inspector already knows the nature and purpose of the experiments to be done, and has usually discussed them with the licensees, whom he knows personally. It must be appreciated that the inspector is not a policeman, and that it would be futile to attempt to oversee every one of a series of similar experiments. The practice has been, in fact, to get to know the scientists concerned and to act so as to *prevent* infringements.

It is interesting to compare all this with any other inspectable activity. One case for which some figures are available is factories, where considerable human hazard may exist; inspection was found to take place about once a year at each establishment. But inspection of pet shops, animal dispensaries, pet-owners, stables, or (perhaps) children's homes, 20 times a year? No. The irony about animal experiment is revealed yet again: the less there is, the more it is inspected; and of all the activities in which suffering may arise, it is that which offers the prospect of less suffering in the future which is most frequently examined.

## The question of public assurance and public responsibility

Much has been said about the 'secrecy' of animal experiment. This arises primarily from the fact that only the inspectorate has effective right of access to experimental work. It is therefore thought that there is no information available to the public about the nature of the experiments conducted beyond what is in the Home Office annual reports. This ignores the fact of the vast body of biomedical literature, in which the whole range of experimental procedures, objectives, and results is available in libraries. What is published will not, of course, give an account of every experiment, since in the nature of research some experiments yield results not in themselves worth publishing, however much they may lead on to later work. The Littlewood Report mentioned a comment by two of their witnesses (both Edinburgh surgeons) that they estimated that only one-quarter of work done was published. This has actually been cited as a representative figure, although it is evidently a trivial sample of information. But there seems no reason to doubt that even if the number of experiments published falls substantially short of the total done, nevertheless the general *nature* of the experiments will be fully represented, simply because communication of results is the point of scientific work. There is, therefore, a mass of information available to the public.

The information is, in fact, sometimes surveyed by members of animal welfare groups. The essential point that emerges when this is done is not improper behaviour by the experimenters, but apparent deficiencies in the law in what it allows. That must always be a matter of personal opinion, but it is instructive to consider some of the cases selected in the field of safety evaluation. These cases were regarded as almost self-evidently revealing an unjustified infliction of suffering.[6] The purpose of the following brief description is to give the reader a chance to make his or her own assessment.

First was a series of experiments injecting nicotine into monkeys, in whom brain electrodes had been placed to record a characteristic change of electrical activity (from slow waves to fast waves) that appears in the transition from sleepiness to alertness—the so-called arousal response. It was found that the change produced by nicotine resembled natural arousal more closely than that produced by caffeine or amphetamine (the active substance in 'pep pills'). The significance of this is in relation to the long-standing puzzle of *why* people smoke, which is probably linked with the ability of nicotine to improve vigilance in boring tasks. (It is a truism in other fields that social control is impossible until root causes are understood.) It also raises the question whether it is or is not justifiable to use animals to try to understand and mitigate the effects of a widespread but voluntary human practice. A similar question arose earlier over the testing of cosmetics. Is society to provide only for the prudent? Or should we recognize that sooner or later we are all imprudent, and provide accordingly?

Second were tests on a new building material for purposes of insulation, *vermiculite*. Asbestos, for which it was hoped vermiculite would be a satisfactory substitute, can produce cancer in those engaged in its production or use. It seems probable that this is in some way related to the fact that the asbestos fibre is exceptionally thin, considerably smaller in diameter than a single cell. Vermiculite also owes its insulating capacity to its very small particle size, although it has a different physical structure; there was therefore good reason to test it. The test was done by injecting either a sample of vermiculite or a sample of asbestos, matched for weight and particle size, into the pleural cavity (the space between the lungs and the ribs) in rats. After two years the two groups, together with some controls injected with saline, were killed and the lungs examined. All three groups remained apparently in perfect health, and put on weight normally. Those treated with asbestos showed the early stages of cancer, but it was still far short of causing any symptoms; this was an important 'positive control', proving that the test would in fact reveal a carcinogenic activity if it was present. Those treated with vermiculite or with saline showed no abnormality. While at first sight it might seem obviously wrong to inject a building material into a rat's lungs, a knowledge of the reasons for the experiment sheds a different light.

Third were some experiments on the arsenical war gas Lewisite and protection against it by British Anti-Lewisite (BAL), already mentioned earlier (p. 159). BAL, discovered at Oxford in the last war, was the first effective antidote to be discovered. As it turned out, Lewisite was not used in the Second World War, although its forerunner in chemical warfare, mustard gas, had been used in the First; some people believe that this was because an effective antidote was found. After the war the experiments done during the war were published, and the capacity of BAL to protect against even really severe eye damage was shown; this required the production of eye damage in control rabbits, and it is these experiments that have often been referred to. It would have been possible not to have done these experiments, but to have waited until human cases arose and then, in uncontrolled conditions, to have tried to find out how best to use BAL and the length of time after exposure over which it was still effective. But it is doubtful whether many people would have preferred this. BAL in fact proved useful during the war in dealing with the toxic effects of another arsenic-containing substance, neoarsphenamine, used in treating syphilis at that time. Subsequently BAL entered the pharmacopoeia since it was found to be an antidote to acute or chronic poisoning with a range of other heavy metals as well as arsenic. The biochemical idea behind its action was very fruitful, and it was the forerunner of other drugs known as 'chelating agents' with a wide range of uses.

Fourth was a proposed fire extinguisher fluid (bromochlorodifluoromethane). It is one of a general chemical class, useful for extinguishing fires because such liquids are themselves non-inflammable and they and their vapour help to exclude oxygen from the burning material. The group as a whole, however, can exert a number of other actions; thus chloroform can produce anaesthesia, carbon tetrachloride liver damage, and hexafluoroethyl ether convulsions. So it was necessary to test whether the new substance was liable to anaesthetize, convulse, or damage the livers of those exposed to it (firemen or other people nearby). In the event it was found, in the dog, to produce convulsions.

Fifth were experiments with a dye known as 'crystal violet' or 'gentian violet'. This was a well-known antiseptic in use for many years, included in the official National Formulary, and used by doctors in a concentration of 0.5 per cent (5 milligrams per cubic centimetre). Its potentiality in causing eye damage was very poorly documented, and not mentioned in standard medical reference works. The experimental work that was criticized showed, among other things, that a very small drop containing one-tenth of a milligram could cause long-lasting eye damage. The work was done on the advice of independent clinicians, to define the hazard in terms of onset, duration, nature, and severity of the effect.

Sixth were experiments on pig conditioning. Here electric shocks were used, delivered by a battery-operated 'goad', already widely used in animal husbandry. The shocks were 'aversive' in the sense that the pigs would not willingly accept them in order to obtain food. The situation was similar to an animal confined by an electric fence, unwilling to risk a shock to reach something beyond the fence. Responses to it included defaecation, urination, and squealing; pigs will, however, do all three simply on being trans-ferred from a pen to a lorry. The work was intended to analyse stress in animal husbandry (for example, in transport or in mixing groups of pigs) and to evaluate under controlled conditions tran-quillizing drugs that could help to reduce such stress.

Seventh were the 'smoking beagles'. The object of these experi-ments was to test a possible substitute for tobacco leaf as a smoking material. The damage done by smoking is well docu-mented, and millions of people smoke. The substances in smoke actually responsible for the damage are not known, although there are a number of likely candidates, including tar, which may well act together. The firm involved had succeeded in developing a material which, according to a number of tests (for instance on the tar obtained from the smoke), offered appreciable advantages over tobacco, and it was envisaged for use mixed with tobacco leaf. It was required of them to test the smoke in some way directly comparable with human use. In the experiments, beagles were trained to smoke for limited periods, but could always remove their heads from the inhalation masks used. The plan was not to wait until gross bodily changes occurred but to kill the animals after a suitable time to study the early changes (if any) produced. A variety of responsible independent visitors all testified to the absence of any signs of distress in the animals. The experiment could have produced valuable information, but in the event, in response to public reaction, it was terminated, and almost no useful information was obtained. The development of safer smoking materials was also virtually stopped in its tracks for some time. There are many questions that arise: some are technical, such as whether the results of such work on the respiratory tract of the dog could be reliably extrapolated to man; but the principal one is a recurrent issue already alluded to, that of the use of animals for the safety evaluation of substances in widespread voluntary human use.

The last case was that of experiments on visual function in kittens. They involved procedures such as sewing up the eyelids of kittens, and would be generally regarded as unacceptable except for a really good reason. That reason was the existence of a condition of functional blindness of the eye known as 'amblyopia': 'functional' means that there is no obvious physical damage, yet the eye does not see. This can arise as a result of the existence of a 'vulnerable period' in the nervous development during early life (at a few years of age). During this period, interference with vision appears able to prevent proper development of the part of the brain concerned with vision. The purpose of the experiments was to analyse the processes involved. The work has already improved the management of eye operations on children (such as for squint), but has a deeper significance for knowledge of the important adaptive changes going on in the brain during childhood. The experiments involved the most meticulous technique, since assessment of any visual changes required behavioural tests which in turn needed healthy and co-operative animals.

The purpose in citing these cases is to show that the public—and particularly the media—also have a responsibility: namely to make sure, in making its own mind up about the justifiability of animal experiment, that it knows or is told the *reasons* for experiments, the background, and the nature of the benefit to knowledge or use that may result. In all these cases the information required was freely available in the literature.

## Practice in other countries

The reader may wonder how animal experiment is dealt with in other countries. It is a huge subject, and I will do no more than indicate some main trends and point to some sources for further information.

First, the Research Defence Society, with the help of the Federation of European Laboratory Animal Science Associations, has compiled a table (here reproduced as Table 14) showing practice in EC countries with respect to ten key features of the British 1986 Act.

It has long been the case that Britain has more rigorous legislation than any other country, but there are now signs of other countries adopting similar policies. The following are some accounts of these developments.

As regards Europe, there are excellent historical accounts up to the eighteenth century generally, and then for France, Italy, Germany, and Sweden, as well as Britain in Rupke (1987). Esling (1981) and Hampson (1989) describe modern European legislation. A good idea of the various attitudes may be gained from the Council of Europe Public Hearing in Strasbourg (8–9 December 1982) where the issues were vigorously debated.[7] The Council for International Organizations of Medical Sciences (CIOMS) has published the proceedings of a conference in Geneva in 1983, which formulated some guiding principles.[8] There is a very useful list of over 20 EEC directives that include animal testing requirements in Purchase 1990; there are likely to be further directives or amendments.

For the United States, there is a useful historical survey by Sechzer (1981), and she organized a wide-ranging symposium in 1983, dealing largely with the issues as seen from an American perspective.[9] Under her chairmanship, the New York Academy of Sciences has put forward some principles and guidelines (1988). Morrison, in on 'Legislation and Practice in the United States', has a useful Table 3.1 listing all the federal agencies involved with

**Table 14.** Laboratory animal welfare: comparison of EC countries with respect to ten key features of the British 1986 Act

| | UK | B | D | E | F | G | WG | I | P | S | H |
|---|---|---|---|---|---|---|---|---|---|---|---|
| Project licences for experiments | ● | ● | | | | ● | | | | | ● |
| Personal licences for qualified persons | ● | | ● | ● | ● | | ● | ● | | ● | |
| Controls on pain and suffering | ● | | ● | ● | ● | ● | ● | ● | | ● | ● |
| Cost–benefit analysis linking pain to purpose of project | ● | | | | | | ● | | | | |
| Requirement to use available alternatives | ● | ● | ● | ● | ● | | ● | | | | ● |
| Inspection of experimental and breeding facilities | ● | ● | ● | ● | ● | | ● | ● | | | ● |
| Provision for daily care of animals | ● | ● | | | | | ● | | | ● | ● |
| Controls on re-use of animals | ● | | | | ● | | | | | | |
| Requirements to use registered breeders | ● | ● | | | | | ● | | | | ● |
| Independent monitoring committee | ● | | ● | ● | | ● | | ● | | | ● |

*Key*: UK = United Kingdom; B = Belgium; D = Denmark; E = Eire; F = France; G = Greece; WG = West Germany; I = Italy; P = Portugal; S = Spain; H = Holland.

*Source*: Research Defence Society 1991

control of toxic substances, and Fox and McGiffin have published an interesting paper entitled 'Live-animal science projects in US schools'.[10] An important report is that on the animal rights movement in the United States by Harvard's Office of Government and Community Affairs (1982); it brings out clearly the number of antivivisectionist organizations and the remarkable financial resources of some of them.

There is an active debate, which is reflected in the pages of journals such as *Science* and *Nature*, which cannot be summarized, since it is so varied and still in progress. The federal structure of the US means that any uniformity depends more on grant-giving bodies than national legislation. Perhaps a resolution adopted in 1990 by the Board and Council of the American Association for the Advancement of Science (AAAS), which is well worth quoting, points to an emerging general view.

Whereas society as a whole, and the scientific community in particular, supports and encourages research that will improve the well-being of humans and animals alike, and that will lead to the cure and prevention of disease; and

Whereas the use of animals has been and continues to be essential not only in applied research with direct clinical applications in humans and animals, but also in research that furthers the understanding of biological processes; and

Whereas the AAAS supports appropriate regulations and adequate funding to promote the welfare of animals in laboratory or field situations and deplores any violations of those regulations; and

Whereas the AAAS deplores harassment of scientists and technical personnel engaged in animal research, as well as destruction of animal laboratory facilities; and

Whereas in order to protect the public, both consumer and medical products must be tested for safety, and such testing may in some cases require the use of animals; and

Whereas the AAAS has long acknowledged the importance and endorsed the use of animal experimentation in promoting human and animal welfare and in advancing scientific knowledge;

Be it resolved that scientists bear several responsibilities for the conduct of research with animals: (1) to treat their subjects with proper care and sensitivity to their pain and discomfort, consistent with the requirements of the particular study and research objectives; (2) to be informed about and adhere to relevant laws and regulations pertaining to animal research; and (3) to communicate respect for animal subjects to employees, students, and colleagues; and

Be it further resolved that the development and use of complementary or alternative research or testing methodologies, such as computer models, tissue or cell cultures, be encouraged where applicable and efficacious; and

Be it further resolved that the use of animals by students can be an important component of science education as long as it is supervised by teachers who are properly trained in the welfare and use of animals in laboratory or field settings and is conducted by institutions capable of providing proper oversight; and

Be it further resolved that scientists support the efforts to improve animal welfare that do not include policies or regulations that would compromise scientific research; and

Be it further resolved that the AAAS encourages its affiliated societies and research institutions to support this resolution.

Finally, there are two important recent developments each with international aspects. First is the formation in the UK of the Research for Health Charities Group (RHCG). This consists of the eight major medical charities, including the Cancer Research charities and the Wellcome Trust with its international range of support (e.g the work on leishmaniasis in Brazil, Figure 14). It will seek to improve public understanding of animal experiment, 'the essential link between test-tube and patient', to counter the harassment of scientists and charity workers, and to correct misinformation. It is significant that the charitable supporters of medical research are now taking action.

The second development is a joint declaration on the use of animals in research signed on behalf of: the British Association for the Advancement of Science, and its Student Section; the Medical Research Council; the Conference of the eighteen medical Royal Colleges and their faculties; the British Medical Association; the General Medical Council; the International Brain Research Organization; and the Society for Neuroscience in Washington, together with Nobel Laureates, past presidents of the Royal Society and many other distinguished scientists.

The declaration is complementary to that by the AAAS, and is also worth quoting.

## Declaration on Animals in Medical Research

In view of the threat to medical research posed by increasingly vocal and violent campaigns for the abolition of animal experimentation, we make the following declaration:

experiments on animals have made an important contribution to advances in medicine and surgery, which have brought major improvements in the health of human beings and animals;

continued research involving animals is essential for the conquest of many unsolved medical problems including cancer, AIDS, other infectious diseases, and genetic, developmental, neurological and psychiatric conditions; much basic research on physiological, pathological and therapeutic processes still requires animal experimentation. Such research has provided and continues to provide the essential foundation for improvements in medical and veterinary knowledge, education and practice;

the scientific community has a duty to explain the aims and methods of its research, and to disseminate information about the benefits derived from animal experimentation;

the comprehensive legislation governing the use of animals in scientific procedures must be strictly adhered to. Those involved must respect animal life, using animals only when essential and as humanely as possible, and they should adopt alternative methods as soon as they are proved to be reliable;

freedom of opinion and discussion on this subject must be safeguarded, but violent attacks on people and property, hostile campaigns against individual scientists, and the use of distorted, inaccurate or misleading evidence should be publicly condemned.

## Summary

This chapter deals with how the various considerations involved in animal work are balanced, as revealed in practice.

1. It is essential that any animal experiments done are of satisfactory standard scientifically. The means of ensuring this are considered by reviewing the criticism to which the scientist is exposed during this work—a critique as severe as that in any walk of life.

2. The trend in animal use, now falling, and the various categories of experimental work are surveyed, using the 1990 Home Office figures (in which the term 'experiment' is now replaced by the term 'procedure' so that about 700,000 animals used in 'production' can be added in). In general, the numbers of the various species used diminish as one moves upwards towards the 'higher' animals (from 1,636,332 mice to 89,845 rabbits, 5,284 monkeys, and 4,392 cats).

3. In assessing whether the numbers used are large or small, it is necessary to appreciate both the number of different categories into which the work falls and the number of observations needed to *establish* any particular conclusion (illustrated also by opinion polls).

4. Procedures wholly under anaesthesia (about 15 per cent) involve as little suffering as the putting down of a domestic pet.

Procedures without anaesthesia totalled about 69 per cent, and with recovery from anaesthesia about 16 per cent. The former are allowed only when anaesthesia would produce a greater stress, except under some very carefully reviewed conditions; the latter are concerned with preliminary preparations before the experiment proper.

5. Despite claims to the contrary, procedures applying substances to the eye (apart from virus propagation for vaccines) totalled only 0.21 per cent, thermal injury 0.42 per cent, and cosmetics and toiletries 0.14 per cent.

6. About 91 per cent of the procedures were concerned with medical and biological research and production. A considerable number are required to meet legislative demands in the UK or overseas.

7. About 0.3 per cent of procedures were used in education, about one per medical student per year, to prepare him for his life's work.

8. To embark on animal experiment, an individual requires stipend, laboratory space, and research expenses (all in short supply), and appropriate licences must be obtained (personal and project). His or her work is then subject to scrutiny by technicians, scientific colleagues, learned societies, journals, and grant-giving bodies. Quite apart from his or her own normal feelings of humanity, these factors are a powerful force against unnecessary animal use.

9. It has been asserted that the majority of scientists engaged in animal experiment become desensitized to animal suffering. It is not explained how those seeking the knowledge to prevent and relieve human and animal suffering are *less* sensitive than those who seek to hinder such work.

10. The inspectoral system is reviewed. The principal sanction is not prosecution but withdrawal of licence from an individual or from premises—both essential to scientific employment. Animal experiment is probably the most inspected human activity there is.

11. The public, too, has a responsibility: namely, to ensure that before taking a decision about the justifiability of particular experiments, it understands the scientific context, purpose, and use.

12. Regulation of animal experiment in other countries is briefly surveyed. British practice remains the most stringent.

# 12

# The future

It has been the theme of this book that there are no 'knock-down' arguments, but rather that animal experimenter and layman alike must balance evidence and arguments over a wide field in order to arrive at accurate and responsible judgement.

It is hoped that the summaries at the end of each chapter will help readers to bring all the points together in their own minds. Rather than recapitulate them, we may now turn to future developments.

There are a good many hopeful signs. Britain has led the way for a hundred years in the care of experimental animals. The issues concerned have recently been debated with great thoroughness.[1] The European Convention at last brings hope that comparable standards will be established elsewhere.[2] If some principles of special importance for future legislation must be named, my choice would be:[3]

(a) That the numbers of experiments done must be returned. This has given a realism to discussion of practice in Britain lacking elsewhere, and provides an invaluable index of trends.

(b) The inspectoral principle offers very considerable and flexible advantages, allows a cadre of independent expertise to be built up, and facilitates uniformity of standards.

(c) Responsibility for the conduct of an experiment must remain with the experimenter him or herself, so that it is the person most closely in contact with the animal who is responsible for its well-being.

(d) However legislation is shaped, it must be broadly drawn, so as to be adaptable to changing circumstances.

Further, there are many directions in which animal suffering is being and will be progressively reduced. In toxicity testing, ways are being found to reduce the number and the severity of tests needed. Both in toxicity testing and in animal experiment generally, advances in technique allow the use of methods which are either 'non-invasive' (in other words, need no surgery) or require less operative interference. More sensitive methods also allow the earlier detection of structural or functional changes. Our knowledge of methods of relieving pain and suffering has greatly improved, and its application in experimental work offers appreciable scope. The standards of scientific work are rising, increasing the amount of knowledge gained from each experiment. There is now a substantial modern literature on animal anaesthesia,[4] on the care and management of laboratory animals,[5] and on the various alternative methods available.[6] Boundaries between the disciplines are becoming less watertight, allowing useful exchange of techniques and ideas.

Behind all this lies the question of the pattern of future developments, in animal care, in scientific advance, in legislation, and in benefit. I believe Lord Justice Moulton's dictum remains true:

> Your duty is to take that line which produces the minimum total pain, and whether the pain is inflicted pain, or whether it is preventable pain that is not prevented, is in my opinion one and the same thing.[7]

Today, to 'pain' one would add 'suffering'. Moulton in that passage did not refer to the perpetuation of ignorance. We would also admit today that there is no simple utilitarian calculus by which we can 'minimize' pain, but that we simply balance all the considerations as best we can. But the objective he stated remains our task, involving the assessment of any inflicted suffering on the one hand, and on the other the assessment of preventable suffering, as judged from the historical record and the suffering we see around us. Animal experimenters must play their part; and earlier chapters have outlined the nature of the work they do, the non-animal alternatives they use, the context in which they work, and the critique and inspection to which they are subject.

## Onward from the three Rs

In 1959 Russell and Burch advanced a programme called the '3 Rs' concerning the use of animals in experimental work: 'replacement' by alternatives, 'reduction' in numbers, and 'refinement' of method. How could one object to such humanitarian objectives? Indeed, this alliterative slogan has become quite fashionable, and many public speakers seek to make suitable allusion to it. But the public should realize that its adherents usually have the objective of *total* and unconditional replacement of animal experiment; that is, in its absolute form it is antivivisectionist.[8] This is not always explained. The tactics of other antivivisectionist movements are sometimes deplored, and the impression of a 'central' position seems to be sought. But the underlying truth is that the ultimate (or 'eventual' or 'top') objective is not the minimizing of human and animal suffering present and future, but the abolition of animal experiment. Recommendations for further elaborate bureaucratic controls multiply, which seems calculated to strangle a scientist's creative impulse at birth. There is little thought as to the research consequences of these controls. The choice is always presented as between 'animal suffering' and 'human benefit', never as between 'animal suffering' and 'future human and animal suffering'. One day, perhaps, a welfare society will point out a region where *more* animal experiment is needed. In the meantime, the 3 Rs have been compared to the Trinity, to quote from an article in March 1990 which, after endorsing the ultimate goal of total replacement, ends:

Thus, the time when thinking Christians are forced by the calendar of events to consider the unfathomable mystery of that other Trinity seems an appropriate moment to reconfirm our additional commitment to our own indivisible three parts of the same whole, the three Rs of Russell & Burch—reduction, refinement and replacement.

The alliterative impulse need not be limited to these three nouns. The pharmacopoeial study described earlier, as well as the decline in the capacity for whole animal research, suggest a need for 'rediscovery', 'reinvention', 'redevelopment'. But there is a more general issue, that of 'responsibility'. Has it been responsible for the media to have focused so often only on experimentally inflicted animal suffering and to have virtually ignored both the historical record of human and animal suffering relieved and all the preventable suffering waiting for relief? Has it been responsible for the animal hooligans, mostly of the healthiest generation this country has ever seen because of past medical research, to harass the experimenters, the animal-breeders, and the industries? Has it been responsible for individuals to say that while, of course, they cannot condone violence, yet if something is not done soon, one must not be surprised at 'direct action'—can condonation go further in a peaceful democracy? The reasons why animal experiment is essential in medical and veterinary research have been explained. Those who unreasonably harass and restrict it now will carry a grave responsibility for the unnecessary ignorance and unnecessary human and animal suffering that will result in the future.

## The depreciation of man?

There is another question to be raised: that of the need for a rather deep change in attitude—to man himself. Sometimes there seems at large something like hatred, for instance in the closing passage of a common sample of Greenpeace literature:

Modern Man has made a rubbish tip of paradise. He has multiplied his numbers to plague proportions, caused the extinction of hundreds of species of animals, ransacked the planet for fuels, and now stands like a brutish infant, gloating over his meteoric rise to ascendancy, on the brink of the final mass extinction and of effectively destroying this oasis of life in the solar system.[9]

Along with this goes the wish to separate man and animal as far as possible, a real apartheid, whatever their interactions and companionships have been in the past.

Commoner, at present, is the repeated depreciation of human significance in the scheme of things. There is a real question as to where this excess of cosmic modesty as to man's status has come from. Partly it must be the repeated rehearsal of the Copernican revolution and man's awareness of the vastnesses around him—although until a comparable being appears somewhere in that universe to join him, it may be more a sense of uniqueness that is felt. Partly it must be from the steady demonstration in the last centuries of the anatomical, physiological, chemical, psychological, evolutionary continuity of animals with man. Darwin's famous passage at the end of his *Descent of Man* puts it eloquently:

Man may be excused for feeling some pride at having risen, though not through his own exertions, to the very summit of the organic scale; and the fact of his having thus risen, instead of being aboriginally placed there, may give him hopes for a still higher destiny in the distant future. But we are not here concerned with hopes or fears, only with the truth as far as our reason allows us to discover it. I have given the evidence to the best of my ability; and we must acknowledge, as it seems to me, that man with all his noble qualities, with sympathy which feels for the most debased, with benevolence which extends not only to other men but to the humblest living creature, with his god-like intellect which has penetrated into the movements and constitution of the solar system—with all these exalted powers—Man still bears in his bodily frame the indelible stamp of his lowly origin.[10]

Thus man has come to be presented as just one more sample of the evolutionary output (though this is not quite Darwin's position). The resultant impact is that of a stern finger being shaken at humanity: 'Now, now, you mustn't be *proud*.' Obediently, the intelligentsia goes out of its way to disclaim or disparage human achievement.

Yet there is something false in all this. The 'indelible stamp' of chisel marks on a sculpture does not make it the equivalent of rubble; nor do brush marks on a painting turn it into random rubbish. Perhaps it is a tendency to focus on thoughts, theories, and emotions that has been misleading. In fact, man's *achievement* has been astonishing, if you simply look at what he has done. One good library, museum, or art gallery is sufficient demonstration; there is nothing remotely comparable to these, or to what lies behind them, to be found in the natural world.

It is therefore profoundly *inaccurate* to depreciate humanity and to equate men with animals, morally, intellectually, or in any other way, just as it is to equate animals with sticks and stones, even if one may not be able to characterize fully the differences and how they arise. The reluctance to admit ignorance may be one of man's greatest weaknesses today, forcing him to take false positions. But that does not alter the facts. A second mental hurdle that he does not take easily is 'suspension of judgement'. Polarization of opinion is too easy; and too readily man must be either a god or a brother to the amoeba. Why should he not be different from both—a being with his own strengths and weaknesses, still to be fully delineated, with his own characteristic role to play?

## Man and animal, trustee and companion

The concept of man's trusteeship for animals flows naturally from all we know. It is interesting that, for instance, J. E. Lovelock, the author of *Gaia*, rejects the idea of trusteeship.[11] In his eyes man is not good enough or responsible enough for the task. 'We are not managers or masters of the earth, we are just shop-stewards, workers chosen, because of our intelligence, as representatives for the others, the rest of life on our planet.' But whatever the image, a responsibility remains, and a difficult one. For it is an earth where, over the course of geological time, evolution's progeny has been of the order of 5–50 billion species, yet only an estimated 5–50 million have survived to today.[12] With such a transformation, can it be man's sole duty that change should cease and the *status quo* simply be conserved?

For man and animals one could look forward to a different relationship, one far from the apartheid mentioned above. Sir Henry Dale used to tell the story of an antivivisectionist friend whom he invited to see an experiment he was doing on the uterine stimulant ergot. The subject was a small bitch, operated on so that a balloon could be placed through the abdominal wall into the uterus to measure its contractions. The bitch was brought in, Dale tapped the bench, and up she jumped and turned on her back ready for the test. When the assay was complete and the bitch was being returned to her kennel, the visitor remarked on her intelligence, and asked if he could have her as a pet when the experiments were complete. 'No', Dale had to reply; 'by the law that you and your fellows have made for us, she must be killed at the end of the experiments.' Many senior physiologists will have had kindred experiences with their experimental animals; and it creates the idea of animals as companions in research as well as outside it. It is strange that animal experiment, alone among the uses of animals, is discussed as a 'necessary evil', when it is the only one from which a general benefit follows. As Kaplan has remarked: 'Who speaks for the sick, for those in pain, for the future?'[13]

Most animal death is to little visible purpose. But as we have seen, more and more ways are being found of screening out and diminishing suffering. Perhaps one can see a future where animal experiment imposes no more on the animal than does domestication, and yet can be seen as providing a new fulfilment for the animal world—a companionship with man in advancing knowledge and, for both, a diminishment of future suffering.

# Appendix 1

## The animal terrorists and the antivivisection movement

In recent years, part of the animal welfare movement has turned to so-called direct action: that is, illegal action and violence in various forms. Medical and veterinary research workers and establishments have shared the attention of these terrorists, along with field sportsmen, anglers, butchers, furriers, and a variety of industrial concerns suspected of being involved in using animals. An outline of the movement's history is therefore needed. The issues themselves, so far as animal experiment is concerned, are dealt with elsewhere. This Appendix will focus on a sample of the events that have taken place, the way the pattern of these has evolved, and their origins, as shown by public records.

There is a difficulty, however, in discerning the pattern accurately. One may notice with this branch of terrorism, as with others, that sometimes there is more than one claimant to a particular action (7.10),[1] and that not all episodes are reported, until they appear, for instance, listed with other offences at a trial. The terrorist seeks to win public kudos among his particular constituency, but his work is less advertised when public revulsion reaches a certain level. Similarly, the victim may either welcome public support and understanding or may not wish to be publicly identified. But there are a good many cross-checks, and the pattern of activity seems clear, if not the full ramifications. For the most part, it is the events that are narrated, with participants left anonymous.

### Origins

The antivivisection movement grew from two roots during the last century. The first was a well-founded impulse towards the welfare of the animals associated with man. There was often a strong religious background to this. The more ancient vegetarian move-

---

[1] The numbers in parentheses in this Appendix refer to a personal file of relevant documentation.

ment, with its wider cultural associations (e.g. as to the 'eating of blood') was sometimes an ally, sometimes not. For instance, Frances Cobbe, that doughty pioneer of the antivivisection movement, did not object to killing animals for food (Rupke 1987, p. 273). The second root was a more specialized reaction to certain animal experiments, especially in the early days of physiology. There was intense debate in the 1870s, which eventually led to the setting up of a Royal Commission on animal experiment in May 1875 (French 1975; Rupke 1987).

As late as 1784, there seems to have been no British legislation of any kind to protect animals (Moss 1961). Individuals, of course, commonly took proper care, and often expressed affection and appreciation of animal companions. But there could be great cruelty, and the methods of slaughter for food could be quite atrocious. Bull-baiting, bear-baiting, cockfighting, and other arranged fights between animals were common. Control or treatment of wild or domestic animals centred on animals as property or as labourers or as predators or vermin to be hunted or otherwise destroyed. Vegetarianism, too, is an old issue. But, as remarked earlier, when much human life could be short and brutal, animal life was not likely to be more highly regarded. Not until 1809 did Lord Erskine succeed in the House of Lords with a Bill to prevent malicious and wanton cruelty to animals, but it was defeated in the Commons.

The first legislative success was that of Richard Martin ('Humanity Dick'), on his second attempt, getting a Bill to prevent improper treatment of cattle through both Houses in 1822. It is sad to record that over the next seven years, ten further efforts by him to extend the protection failed.

Two years later, in 1824 in London, the Society for the Prevention of Cruelty to Animals was founded by The Revd Arthur Broome (a product of Balliol College, Oxford), backed by such as Fowell Buxton, the great anti-slavery campaigner, and the engineering inventor Lewis Gompertz. There were internal controversies as there are today. But the Society grew. Perhaps a letter in 1835 from the Duchess of Kent to say that she and 'Princess Victoria', later to be Queen, would agree to be patronesses marks the clear establishment of a new welfare force, soon to become today's RSPCA.

The general impulse to control cruelty to animals led eventually, at the legislative level, to the Protection of Animals Act of 1911. This consolidated 8 prior Acts (including Martin's), and is still in operation, though subject to around 15 subsequent Acts dealing with special matters. It is a very general Act, concerned not only with the personal *infliction* of any 'unnecessary suffering', but also with 'allowing it'. It deals with 'domestic' animals (i.e. any species which is tame or tameable to serve some purpose for the use of man) and with 'captive' animals (any species including bird, fish, and reptile, which is in captivity or confinement). Wild animals, whether hedgehog or wolf, are outside its range. An important feature is that it is essentially reactive, and action can be taken under the Act only after an offence has been committed. It remains the principal control of cruelty to animals, although the offences have changed in character.

The more particular concern with animals subjected to experiment had its own specific legal outcome in 1876. It was called the 'Cruelty to Animals' Act, and it laid down the conditions to be met before animal experiments could take place. Its first sentence uses the phrase 'experiments *calculated* to inflict pain' (my emphasis). For the medical scientist working on the relief of pain, some of this language must have seemed unjust. Sometimes even today it is unjustly exploited; for example, the description of such a scientist's work as 'cruelty' may be used to convey the idea of actually taking pleasure in suffering, and the phrase 'calculated to inflict pain' to suggest calculating sadism rather than simply that pain may arise. But it needs to be remembered that the Act started as a draft based on a submission to the Home Secretary by the antivivisectionist Victoria Street Society and that a rival Bill in the Commons had proposed total abolition of vivisection. Such language was part of the political compromise. It is interesting to study Figure 1 (on p. 60 of this book) on the supposition that the latter Bill had succeeded.

The Act allowed pain to be inflicted within limits if it was to advance knowledge, to save life, or to alleviate suffering. It is rarely pointed out that it also differed from the 1911 Act in being *preventive*, laying down conditions to be met before experiment could be undertaken—for example, the licensing of investigator and premises. Thus, even if no actual suffering occurred during an experiment, an offence could still have been committed if the appropriate licence had not been approved or if the Home Office had not licensed the premises. The same principle applies to the successor 1986 Act.

At this time children were still outside the law. It is interesting that a letter in 1870 urged the Society to include children in its remit. But it needed a stimulus from the USA before the National Society for the Prevention of Cruelty to Children was formed in 1884. In the light of current debate, the story is relevant. It is said that an American lady, disturbed by a report of child cruelty, went to Mr Bergh of the American SPCA to report that 'there is a little animal suffering from the unkind treatment of a bad woman'. He promised to look into it. He still kept his promise when told it was a child. The SPCA investigated, and brought the case to court, arguing successfully that the child was an animal. Many similar cases then followed, and this led to the formation of the New York Society for the Prevention of Cruelty to Children. This was then followed by the NSPCC in London—60 years after the RSPCA.

The debate leading up to the Cruelty to Animals Act gathered together antivivisection opinion, some of which proved to be dissatisfied with the RSPCA's attitude. So in 1875 the 'Victoria Street Society for Protection of Animals Liable to Vivisection' was formed, under the guidance of Lord Shaftesbury and the colourful Frances Cobbe (1822–1904), who had campaigned previously in Italy. Some time later, a division arose between those who were disappointed by the Act and wished to press for immediate total abolition of animal experiment and others, gradualists, led by Stephen Coleridge. Coleridge replaced the aging Miss Cobbe; and under his leadership, the Victoria Street Society changed its name to the 'National Anti-Vivisection Society' (NAVS) of today, to correspond to restrictionist rather than abolitionist aims. Undaunted, Frances Cobbe then founded the British Union for the Abolition of Vivisection (BUAV) in 1898.

Although welfare organizations have multiplied abundantly (there are said to be around 300 today registered as charities), it is striking that the three societies so far named have formed the principal background, maintaining roughly the same mutual relations ever since the turn of the century. At any particular time, of course, the tone of a society reflects its membership; activists and conservatives come and go. But today it is still the RSPCA that represents the broadest approach to animal welfare, doing invaluable work in dealing with general cruelty to animals, as well as seeking to get rid of unnecessary animal experiment. The NAVS favours a gradualist, non-violent approach to abolition; and the BUAV (recently somewhat activated) is strongly abolitionist.

This brings us to an important change: the emergence of violence. Euphemistically called 'direct action', this was initiated on the grounds that the usual methods of argument and demonstration have failed and that stronger methods must therefore be used. The initial claim was that violence against people was completely ruled out; the objective was simply to attack property so as to cause economic loss to those whose activities were regarded as infringing animals' rights. It is instructive to follow the course taken by subsequent events.

### The new Band of Mercy

To go back for a moment, one of the many activities associated with the early RSPCA and the welfare movement had been educational, focusing on the need to teach children kindness to animals. In 1875 a certain Mrs Smithies founded a 'Band of Mercy for Promoting Kindness to Animals' at Wood Green, believing that 'the teaching of children to be kind and merciful to God's lower creatures is preparing the way for the gospel of Christ'. (Her husband was a great temperance advocate, proprietor of the *British Workman* and later founder of *Animal World*) In due course there were other Bands, a journal and a medal for members, comparable to the children's badge issued since 1856 by the RSPCA. Indeed, the movement was probably rather overshadowed by the RSPCA. The latter had, through its Ladies Committee, itself inaugurated many educational efforts; and in 1883, the Band of Mercy movement was absorbed by it.

In 1973, it was the name of this gentle body that was taken by the new generation of animal activists. They themselves had split away from the groups protesting against hunting and seal-culling, deciding to replace placard protest by direct vandalizing of hunting grounds and the burning of sealing boats. An early extension of activity by the new Band of Mercy was arson at Hoechst's laboratories in December 1973. Considerable damage was done. A notice stated that it was 'an effort to prevent the torture and murder of our animal brothers and sisters by evil experiments'. It is ironic that Hoechst was a major supplier of vaccines to treat animal diseases; today they manufacture at least 10 of the vaccines listed in Table 7 on p. 97.

These were not in fact the first attacks by activists. As early as 1968, activists had released cage-birds being used by Professor Thorpe at Cambridge for his work on bird-song. The birds, chaffinches and budgerigars, were, of course, soon taken by jackdaws.

The attacks became more varied: on animal-breeders and animal-suppliers, as well as on laboratories. Principal actions were starting fires, setting fire to vans, slashing tyres and spraying paint, and destroying property. In time, two leaders were arrested, and charges followed in August 1974. Prison sentences of 3 years were awarded in February 1975; release on parole took place after 1 year.

Alongside the start of these events, it is interesting to note the report of a protest meeting in Oxford in September 1973, organized by the NAVS and addressed by Richard Ryder (clinical psychologist in Oxford and author of *Victims of Science* (1975)) and Peter Singer (philosophy lecturer at University College Oxford and author of *Animal Liberation* (1975)). These widely read books (with some preliminary essays) and the activism were of roughly the same date.

## The Animal Liberation Front

In October 1975, the then chairman of the Band of Mercy disclaimed an attack on a meat factory: 'In the past arson was used and two men imprisoned. But it is not now our policy to use such methods.' This did not last. Release of the Band of Mercy offenders in 1976 was accompanied by the formation of a new organization, the 'Animal Liberation Front' (ALF), and by renewed activity of the earlier type, with additions. It still claimed, however, to be against any form of personal violence and to wish only to inflict financial damage on those concerned with exploiting animals. To the 'non-violence' of arson and van-daubing and tyre-slashing were now added theft of animals (such as pregnant beagles, mice, rabbits, cats (1.4); the use of smoke-bombs and flashes; attacks on offices, with theft of files; attacks on personal homes such as that of a member of the Home Secretary's Advisory Committee and that of an animal technician who appeared on television; an arson attack on a medical school; fire bomb and other threats; and letter or telephone harassment.

This phase up to about 1982 may be described as moving from *impersonal* attack on property towards *personal* threat and attack on individual homes, together with animal theft. Difficulty in looking after or disposing of the stolen animals was soon recognized (3.11); indeed, it would be a good subject for investigative journalism to find out objectively what did happen to them all.

Some modes of thought were revealed. Among the less attractive examples were a letter to one scientist and his wife, who were expecting a child, 'hoping that the infant would be born blind' (3.28). Another wrote to the author: 'May you rot in hell or even return in a "second life" as a Draize-tested rabbit.' A federation of welfare societies declared that 'the indifference of heartless politicians is driving our young people into this kind of action' (4.1).

In 1977 a raid on a drug firm and theft of 125 mice (3.11) provided evidence for a further conviction of two ALF members, with a sentence of 12 months, of which 8 months were served. The ALF appeared to have become a loosely knit organization, made up of a number of largely independent groups, using relatively simple means. Thus police action against one group could be successful without preventing continuing action by others.

### 1982–1988: letter-bombs: contamination threats: coshing: car bombs

In continuing the narrative, there is a difficulty: either it becomes disjointed or it is forced into a historical mould that may be wrong. Accordingly, what follows may be rather disjointed, but the incidents reported are arranged roughly chronologically; and while it is far from a complete account, an attempt has been made to include samples of the different types of activity.

The year 1982 was marked by a new and serious escalation: namely, the actual sending of *letter-bombs*. The Prime Minister and other leading politicians were the first targets, with other individuals becoming targets later (7.10). The Downing Street office-worker who opened the bomb had his hair burnt and his left cheek scorched; fortunately, he was wearing spectacles. The operation was claimed by a new body, the 'Animal Rights Militia' (ARM), among others. A few months later, similar ARM attacks were made on another government minister, a high commissioner, a medical school, and a veterinary school (7.22). The ARM reappeared later with a reported threat to break the hands of scientists who work with animals, 'so they won't be able to torture any more' (30).

For a while, it was said that this could not be anything to do with the ALF and that 'only a lunatic would do this' (7.10). The ARM seemed not to have been heard of. It was even said to be an invention by the opposition to discredit the animal rights movement (7.22). As time passed, however, the ALF proved willing to publish ARM threats of plans to attack 'real animal abusers' (76). Later (1986), ALF distributed a booklet of detailed instructions on making bombs, as well as a range of methods of harassment (91). Despite the non-violence claim of the ALF leaders, one of them appeared in a film saying that 'if vivisection continued, one of these professors will get shot on the doorstep' (7.8; 7.5). The interaction seems sufficiently clear. One view expressed has been that the ARM is just a name giving an umbrella to individual activists of the ALF.

In other respects, too, there were changes. One novelty was the release of around 3,000 *mink* from a mink farm (20). Most of the animals stayed where they were, despite 'release', but a number escaped to run wild and slaughter ducks, chickens, pheasants, and each other.

There is a useful survey of ALF activity in the first half of 1983, by a sympathizer using their files (Duffy 1984). This helps to form an overall picture. Objects of attack included animal–dealers, animal houses, battery egg farms, butchers' shops, equipment manufacturers, furriers and fur traders, hunts, laboratories (agricultural and medical), slaughter houses; a supermarket freezer department, and a taxidermist. Methods used included breaking windows (one method was by catapult, developed by the ALF), stealing or releasing animals, damaging cars, supergluing locks, painting slogans, and damaging water-pipes and generators. Interestingly, the letter-bombs sent to Mr Peter Walker and to two universities are not listed. The list shows that animal experiment was an important focus, but not the only one.

Another noticeable feature was that some laboratory attacks came to involve considerable numbers of people, and arrests of up to 50 people were not unknown (20.2).

It should also be recorded that in March 1983 Peter Singer was howled down at a meeting because of his absolute repudiation of violence and letter-bombs (7.20).

There was a further escalation. One notable instance involved *physical violence to individuals*. This was a multiple raid on two laboratories, a kennels, and a flat (29; 64; 72). Scores of people were active. Of the victims, one man was coshed, while a woman, her son, and her daughter-in-law, were trussed up with ropes round their necks and dumped.

Other developments included *fire-bomb attacks on private homes* (28), and telephone threats and daubing of animal blood and paint at a Government Minister's home (22.1; 27). Broken glass was spread on the Coventry City pitch on the day for playing Newcastle United, whose manager had been shown in field sports on Channel 4 (30). A further bizarre event, reputedly by the 'Hunt Retribution Squad', was an attempt to dig up the grave of the Duke of Beaufort, who had died earlier in the year, aged 83 (37).

The welfare movement at large usually, but not always, deplored such actions. Thus, an ALF leader was expelled from the BUAV headquarters for writing in a newsletter that the ALF 'disdained violence for tactical reasons, not because we love the scum who brutally exploit animals' (24 April 1984). But the BUAV was clearly divided; a year later (1985) it voted *not* to condemn extremists who injure or kill (60). The field sport activists were likewise split over the question of direct action against anglers (21.9; 21.6; 24.8; 32); some action against *anglers* followed later, but it was also noted that about 3.5 million people angle, many of whom vote Labour.

Another novelty, for this country (already familiar in USA (86)), was *contamination threats*. Late 1984 ushered in a spate of claims to have contaminated various home products marketed by disapproved stores. Among such claims were that a baby oil had been spiked, mercury put into turkeys, Mars bars poisoned, bleach added to shampoo, and syringes containing rat poison placed in supermarkets. The threats seem to have been fictitious, a means of causing financial loss to certain firms, without regard to the distress caused to the customers involved—largely women and children. It was reported that in an unpublished letter to *The Times* following the Mars bar scare, 14 animal rights organizations had dissociated themselves from such actions. This did not, however, stop recurrence over the next 7 years (32; 33; 35; 30; 55; 87; 88; 100; 116; 141).

Fire-bombs continued, personally directed, with a public threat that 'more than a 100 scientists were singled out for attention in the coming months'. A serious event, claimed by the ARM, was placing *potentially lethal bombs* under the cars of four scientists, apparently with warnings to Scotland Yard (78/9). New types of activity included the theft of 'pathogen-free' animals from a laboratory. These are animals specially bred to be free from a prescribed range of organisms, so that their physiology provides an especially clean baseline, thereby reducing the number of animals needed (63). There was a raid on a public health laboratory, during which diphtheria culture plates were smashed (63). Among animals involved in one raid were some monkeys in quarantine for rabies and simian herpes (59).

*Finance*

1987–8 saw important groups of the ALF charged by the police, and sentences ranging from 4 to 10 years resulted (91; 93; 97; 99; 104; 104). From the Press reports of these cases, something of the economics of antivivisection and terrorism began to emerge. One serious raid, in which physical violence had been used, had been financed, according to the judge, by a £2,000 donation from the BUAV. Laundering of £23,000 for the ALF was reported in another case (104). Documents from the NAVS revealed its dealing with an offer by the ALF support group of documents, video, and photos for £4,000. The NAVS also paid the Central Animal Liberation League £5,000 for similar material (99). How extensive such use of raids to generate cash may be is not likely to be known in full for many years. It appears that active supporters of the ALF pay £2 per week, and there are street donations and pamphlets. As to more orthodox financing, an earlier feature in *The Times* gave the resources of the RSPCA as £9 million, BUAV as £450,000, and Animal Aid as £200,000 (55).

## Nomenclature

A provisional note on the range of organizations involved may be helpful. The author is aware of at least 46, but this is certainly a substantial underestimate. Some are general, some very specialized as to region, activity, or species. The names are usually formed by adding the word 'animal' or 'animals' to some operational word such as 'aid', 'advocate', 'activist', 'defence', 'liberation', 'protection', 'rights', 'sanctuary', or 'welfare', together with a collective word such as 'front', 'group', 'coordinating', 'league', or 'mobilization'. Of those groups claiming non-violence, the Animal Aid groups and Animal Rights groups seem fairly established. Among the violent groups, ALF, ARM, and the various Liberation leagues are notable. They all use a great deal of common material in their leaflets and publications. One senses a good deal of interchange of membership. Apart from any specialized interest, there seems little difference between them, except for the attitude to the use of violence.

### The Bristol and Porton Down bombs

But the most serious developments were to come. In February 1989, a bomb, estimated at 5 lbs, destroyed the fourth floor of the administrative block at Bristol University (106). There had been a phone warning at 11.30 a.m. that a bomb would go off in 30 minutes. Evacuation and search followed, but nothing was found, and the explosion did not take place as warned. It was therefore regarded as a hoax. The explosion in fact occurred the following midnight—fortunately when no one was there. The ALF has published details of bomb-making, mentioned above, from which it is clear that control as to which 12-hour period was concerned was faulty. It was not the first attack at Bristol; there had been threats, supergluing, and a letter-bomb previously (106). This was the *first major bomb* in the country.

Two months later, three men entered a fur shop in London, aimed guns at the assistants, telling them to lie on the ground, threw three smoke-bombs to damage the furs, and made off in a waiting car. This was the *first use of a gun* (107).

In June 1990, at Porton Down, the ALF planted a car bomb arranged to go off as the driver, a government scientist giving veterinary advice, left for work. There was a huge explosion, with flames at the back of the car, and the windows blew out. The door frames buckled, but she was able to escape through the driver's window, shocked and singed.

A few days later, in Bristol, a bomb was placed under the car of a university scientist. When it went off, the explosion blew a 2 foot hole with 9-inch upward spikes in the chassis under the passenger seat. The driver escaped with a cut nose and temporary deafness. But the car happened to be passing a father with a child in a push-chair at the time. The father was shielded by a parked car from the blast, but not the small boy; a piece of metal lodged near his spine, a finger was partly severed, with blood pouring from the hand, and his back was lacerated (124-5-6). The rapid response of the scientist resulted in the prompt arrival of an ambulance, and the skill of the surgeons has meant that the child has few ill effects.

## Conclusions

We may leave the narrative of events at this point, which one hopes is the nadir, and turn to some assessment of 20 years of activism.

The terrorists have made simple mistakes: for example, mis-identifying targets, such as a timber firm making roof trusses, not broiler houses (101), and a naval research establishment concerned with research on paint, not animals (101); attacking the homes of individuals wholly unconnected with animal experiment, including a highly distinguished epidemiologist (4.6v); allowing acid (the highly toxic hydrofluoric acid, stolen (112.1) and used for etching slogans on butchers' windows (91)) or oil from a blown-up reservoir to leak into surface water (97). There has sometimes been a pettiness, such as digging up the grave of the aged and 10-months dead Duke of Beaufort (37), damage to small family butcher's shops, threats to an animal sanctuary (83), and a raid on a Barnardo's Home to remove some rabbits and a duckling from oppression by children (87). But probably the greatest strategic error has been the latest bombings, lethal bombs with either no pretence at warning or only a futile one. The combination of such actions with professions of non-violence has ceased to convince.

The account above illustrates how the violence has escalated over the years. The impact of this on the welfare societies has been complex. Animal welfare opinion often showed itself torn between apprehension of the effect on support for their own efforts and a sympathy, getting as near to condonation as possible (1.6; 4.9; 20.2; 33; 14 March 1980). Whether violent or non-violent, there have been many organizations in the field competing for attention and funds. At the beginning, rejection of violence was pretty general. But as time has gone on, a more ambiguous tone has developed (1.6; 4.9; 33; 14 March 1980). For instance, one organization, Animal Aid, produced leaflets naming particular scientists. Shortly after, it was these people who received threatening phone calls and warning of militant action. Animal Aid disclaimed the calls, and professed to be opposed to any form of violence, verbal or physical (3.21). Yet it was not much later, as the leafleting and debate went on, that the same organizer welcomed being able to use photographs of work inside laboratories following the 'successful' break-in at Babraham, near Cambridge (4.3). The payments by the welfare societies for material from raids, referred to above, provide another instance (99; 104). But as the violence became more and more personal and severe, some revulsion seems to have been felt.

It is possible to overestimate the significance of these events, since many accounts rehearse earlier episodes. They have certainly had a high nuisance value, and have been much more than 'nuisance' to some individuals. It must be remembered, too, that a new violence came in many other areas: Vietnam, nuclear weapons and nuclear energy, picketing, taxation, race relations, inner city activity, sport, regional independence. The animal terrorists fit well into the class that Bernard Levin has aptly called 'single issue fanatics'. Not all the comment genuinely faced the implications of violence in a democracy. On balance, the terrorism may not have made much difference to changes in practice and legislation; quieter measures aimed at reform in regard to animals have always had a good hearing in England.

In such issues, there is a close interaction with the media. It was notable from the earliest days of the ALF how close their relations were with the Press and media, the latter being used, as with the IRA, as the primary medium for the terrorists' propaganda. The media have not always been even-handed, either in their attitude to the terrorists or in their approach to animal experiment. They have been willing to use illegally obtained material (ICI). They were willing to allow terrorists to appear on screen ('The Animals

Film' 7.3; 21.1). Complaints against the BBC by a university (21.5) and against a newspaper by a neurophysiologist (101) have been upheld. One firm, subject of a vicious animal rights attack, won £45,000 from Central ITV for libel in a programme trying to discredit animal experiment. Most conspicuously, the media have almost totally failed to support the research worker by elucidating how, and how greatly, animal experiments have led to medical and veterinary benefits (7.1/82). A notorious example is a programme whose very title, 'Rabbits don't cry', is a falsehood (7.1).

But there is a price to be paid for media attention. To retain it, activity has to be continually scaled up—Who would make a great fuss about a letter-bomb now? It is not unlikely that the violence has now reached a point at which escalation to secure attention is still required, yet could now arouse only revulsion. At such a point, the terrorism has failed, and reason can be reinstated.

# Appendix 2

## Protection of Animals Act, Clauses 1 and 15

An Act to consolidate, amend, and extend certain enactments relating to Animals and to Knackers; and to make further provision with respect thereto.

[18th August 1911]

Be it enacted by the King's most Excellent Majesty, by and with the advice and consent of the Lords Spiritual and Temporal, and Commons, in this present Parliament assembled, and by the authority of the same, as follows:

### Clause 1. [Offences of cruelty].

1. If any person—

(a) shall cruelly beat, kick, ill-treat, over-ride, over-drive, over-load, torture, infuriate, or terrify any animal, or shall cause or procure, or, being the owner, permit any animal to be so used, or shall, by wantonly or unreasonably doing or omitting to do any act, or causing or procuring the commission or omission of any act, cause any unnecessary suffering, or, being the owner, permit any unnecessary suffering to be so caused to any animal; or

(b) shall convey or carry, or cause or procure, or, being the owner, permit to be conveyed or carried, any animal in such manner or position as to cause that animal any unnecessary suffering; or

(c) shall cause, procure, or assist at the fighting or baiting of any animal; or shall keep, use, manage, or act or assist in the management of, any premises or place for the purpose, or partly for the purpose, of fighting or baiting any animal, or shall permit any premises or place to be so kept, managed, or used, or shall receive, or cause or procure any person to receive, money for the admission of any person to such premises or place; or

(d) shall wilfully, without any reasonable cause or excuse, administer, or cause or procure, or being the owner permit, such administration of, any poisonous or injurious drug or substance to any animal, or shall wilfully, without any reasonable cause or excuse, cause any such substance to be taken by any animal; or

(e) shall subject, or cause or procure, or being the owner permit, to be subjected, any animal to any operation which is performed without due care and humanity;

such person shall be guilty of an offence of cruelty within the meaning of this Act, and shall be liable upon summary conviction to a fine not exceeding twenty-five pounds, or alternatively, or in addition thereto, to be imprisoned, with or without hard labour, for any term not exceeding six months.

2. For the purposes of this section, an owner shall be deemed to have permitted cruelty within the meaning of this Act if he shall have failed to exercise reasonable care and supervision in respect of the protection of the animal therefrom:

Provided that, where an owner is convicted of permitting cruelty within the meaning of this Act by reason only of his having failed to exercise such care and supervision, he shall not be liable to imprisonment without the option of a fine.

3. Nothing in this section shall render illegal any act lawfully done under the Cruelty to Animals Act, 1876,[1] or shall apply—

(a) to the commission or omission of any act in the course of the destruction, or the preparation for destruction, of any animal as food for mankind, unless such destruction or such preparation was accompanied by the infliction of unnecessary suffering; or

(b) to the coursing or hunting of any captive animal, unless such animal is liberated in an injured, mutilated, or exhausted condition; but a captive animal shall not, for the purposes of this section, be deemed to be coursed or hunted before it is liberated for the purpose of being coursed or hunted, or after it has been recaptured, or if it is under control.

[1] 39 & 40 Vict. c. 77.

Clause 15 [Definitions].

In this Act, except the context otherwise requires, or it is otherwise expressly provided—

(a) the expression 'animal' means any domestic or captive animal;

(b) the expression 'domestic animal' means any horse, ass, mule, bull, sheep, pig, goat, dog, cat, or fowl, or any other animal of whatsoever kind or species, and whether a quadruped or not which is tame or which has been or is being sufficiently tamed to serve some purpose for the use of man;

(c) the expression 'captive animal' means any animal (not being a domestic animal) of whatsoever kind or species, and whether a quadruped or not, including any bird, fish, or reptile, which is in captivity, or confinement, or which is maimed, pinioned, or subjected to any appliance or contrivance for the purpose of hindering or preventing its escape from captivity or confinement;

(d) the expression 'horse' includes any mare, gelding, pony, foal, colt, filly, or stallion; and the expression 'bull' includes any cow, bullock, heifer, calf, steer, or ox, and the expression 'sheep' includes any lamb, ewe, or ram; and the expression 'pig' includes any boar, hog, or sow; and the expression 'goat' includes a kid; and the expression 'dog' includes any bitch, sapling, or puppy; and the expression 'cat' includes a kitten; and the expression 'fowl' includes any cock, hen, chicken, capon, turkey, goose, gander, duck, drake, guinea-fowl, peacock, peahen, swan, or pigeon;

(e) the expression 'knacker' means a person whose trade or business it is to kill any cattle not killed for the purpose of the flesh being used as butcher's meat, and the expression 'knacker's yard' means any building or place used for the purpose, or partly for the purpose, of such trade or business, and the expression 'cattle' includes any horse, ass, mule, bull, sheep, goat, or pig;

(f) the expression 'pound', used in relation to the impounding or confining of animals, includes any receptacle of a like nature.

# Appendix 3
## Albert Schweitzer and Reverence for Life

At one time Schweitzer and his ethical principle, 'reverence for life', were much quoted, not always with understanding of the latter's meaning. But he fell out of fashion, particularly for what was regarded as his paternalist attitude to the black races. There is a good biography (G. Seaver, 5th edn., 1955) which gives an account of his liberal theology, his musical ability, organ building and Bach scholarship, and his medical missionary work. He was not a vegetarian, had a gun for shooting snakes but little else, and would lift a worm from the road to safety on to the grass. But he is best heard in his own invariably thoughtful words. I have attempted, therefore, after a brief survey of how his thought developed, to give some flavour of the man through short extracts from his writings.

The full development of Schweitzer's ethical outlook is to be found in his *Civilization and Ethics* (1923), based on lectures in Oxford in 1922. His review of past thought brought him to recognize a need for man to recover an adequate world outlook, an outlook that must be rational, based on a reliance on truth that will not be abandoned in the face of mysticism, optimistic in affirming the value of existence and human life (he called this the 'will-to-life'), and with an impulse to betterment which extends to others as well as the self.

But the idea of 'reverence for life' did not arise just as a general expression of benevolence. Instead, it came to him as a means of synthesizing two deep lines of thought, both of them discussed in terms unfamiliar to us today.

First was the idealistic search for the 'highest possible', leading to mysticism, passivity, resignation from the world, and absorption into an Absolute, particularly as exemplified by the Eastern religions he had studied. This he called the 'ethic of self-fulfilment'. It had universality, but did not provide for action in the real world.

Second was the utilitarianism of Bentham and Adam Smith, built on an aggregation of individual ethics and sharply distinguished from the ethics of Spencer and others later, with its move away from individual to social ethical decision. This he called the 'ethic of self-sacrifice'. But he found that such utilitarianism 'is occupied mainly with the self-sacrifice of one man for another or for human society', and lacks the universality found in a search for an Absolute. He could not see how to bring these lines of thought together.

One of the extracts below, written much later, describes how his insight, 'reverence for life', came to him during a river journey through the tropical forest, where man can be seen so clearly as part of a profuse struggle of life with life. The resolution of his problem by the concept of reverence for every living thing can be illustrated by two extracts from *Civilization and Ethics*:

Self-sacrifice must no longer be directed only toward the individual and the community, but must somehow be applied to all the life which comes into being in the world. (p. 238)

The fundamental principle of ethics ... is expressed by the following formula: *self-sacrifice for the sake of other life motivated by reverence for life itself*. (p. 250; emphasis original)

The more one reads Schweitzer, the more one finds that he's worth quoting. The extracts that follow, however, are restricted to what he says were important moments in his life, in roughly chronological order, and some examples of what his ideas meant in practice. Explanatory background is in square brackets.

### From *Memoirs of Childhood and Youth* (1924)

[An account of his first feeling ashamed, while still in petticoats. He was on a stool while his father worked on the beehives. Suddenly] a pretty little creature settled on my hand, and I watched with delight as it crawled about. [Then it stung him and made him shriek. He remarks that] It had a good right to be angry. [Everyone pitied him and he went on crying with much satisfaction, till he suddenly noticed that the pain had disappeared]. My conscience told me to stop, but in order to be interesting a bit longer I went on with my lamentations, so getting a lot more comforting. However, this made me feel such a little rogue that I was miserable over it all the rest of the day. How often in after life, has this experience warned me against making too much of whatever has happened to me. (p. 10)

As far back as I can remember I was saddened by the amount of misery I saw in the world around me. (39)

[A friend invited him to join in shooting birds with a catapult, which he accepted for fear of being laughed at. But just as they were aiming at their target, some church bells rang, like 'a voice from heaven', and he shooed the birds away.] This early influence upon me of the commandment not to kill or torture other creatures is the great experience of my childhood and youth. (42)

The thought that I had been granted such a special happy youth was ever in my mind; I felt it even as something oppressive, and ever more clearly there presented itself to me the question whether this happiness was a thing that I might accept as a matter of course. Here, then, was the second great experience of my life, viz. this question about the right to happiness. (81)

When I was one and twenty, while still a student, I resolved to devote my life till I was thirty to the office of preacher, to science, and to music. If by that time I should have done what I hoped in science and music, I would take a path of immediate service as man to my fellow men.... Finally a chain of circumstance pointed out to me the road which led to sufferers from leprosy and sleeping sickness in Africa. (82–3)

All ordinary violence produces its own limitations, for it calls forth an answering violence which sooner or later becomes its equal or its superior. But kindness works simply and perseveringly; it produces no strained relations which prejudice its working; strained relations which already exist it relaxes. Mistrust and misunderstanding it puts to flight, and it strengthens itself by calling forth answering kindness. Hence it is the furthest-reaching and most effective of all forces. (103)

From *On the Edge of the Primeval Forest* (1922)

[An account of his medical work at Lamborene, on the river Ogowe in Gaboon, West Africa, just south of the equator. Scabies, strangulated hernia, sleeping sickness, leprosy, malaria, elephantiasis, osteomyelitis, tropical dysentery, gross pyorrhoea, gross tropical ulcers, heart disease, were the main problems. Necessarily simple hospital facilities went with up-to-date drugs and technique.] How shall I sum up the resulting experience of these four and a half years? On the whole it has confirmed my view of the considerations which drew me from the world of learning and art to the primeval forest. 'The natives who live in the bosom of Nature are never so ill as we are, and do not feel pain so much.' That is what my friends used to say to me, to try to keep me at home, but I have come to see that such statements are not true. Out here prevail most of the diseases which we know in Europe, and several of them—those hideous ones, I mean, which we brought here—produce, if possible, more misery than they produce amongst us. And the child of nature feels them as we do, for to be human is to be subject to the power of that terrible lord whose name is Pain. (pp. 170–1)

## From *My Life and Thought* (1933)

I saw, indeed, the conception needed before me, but I could not grasp it and give it expression.

While in this mental condition I had to undertake a longish journey on the river. I was staying with my wife on the coast at Cape Lopez for the sake of her health—it was in September 1915—when I was summoned to visit Madame Pelot, the ailing wife of a missionary, at N'Gômô, about 160 miles upstream. The only means of conveyance I could find was a small steamer, towing an overladen barge, which was on the point of starting. Except myself, there were only natives on board, but among them was Emil Ogouma, my friend from Lambaréné. Since I had been in too much of a hurry to provide myself with enough food for the journey, they let me share the contents of their cooking-pot. Slowly we crept upstream, laboriously feeling—it was the dry season—for the channels between the sandbanks. Lost in thought I sat on the deck of the barge, struggling to find the elementary and universal conception of the ethical which I had not discovered in any philosophy. Sheet after sheet I covered with disconnected sentences, merely to keep myself concentrated on the problem. Late on the third day, at the very moment when, at sunset, we were making our way through a herd of hippopotamuses, there flashed upon my mind, unforeseen and unsought, the phrase, 'Reverence for Life'. The iron door had yielded: the path in the thicket had become visible. Now I had found my way to the idea in which world- and life-affirmation and ethics are contained side by side. (p. 185)

To the man who is truly ethical all life is sacred, including that which from the human point of view seems lower in the scale. He makes distinctions only as each case comes before him, and under the pressure of necessity; as, for example, when it falls to him to decide which of two lives he must sacrifice in order to preserve the other.

I rejoice over the new remedies for sleeping sickness, which enable me to preserve life, whereas I had previously to watch a painful disease. But every time I have under the microscope the germs which cause disease, I cannot but reflect that I have to sacrifice this life in order to save other life.

I buy from natives a young fish-eagle, which they have caught on a sand-bank, in order to rescue it from their cruel hands. But now I have to decide whether I will let it starve, or kill every day a number of small fishes, in order to keep it alive. I decide on the latter course, but every day I feel it hard that this life must be sacrificed for the other on my responsibility. (271)

Schweitzer's writing thus reveals a deep enjoyment of the natural world, a rejection of unqualified rights without duties, an absolute rejection of violence, and a rule to preserve life and not to kill any living thing (down to bacteria) unless there is a necessity. He did not disguise the moral conflict that results: 'One existence survives at the expense of another of which it yet knows nothing.' But evolution has enabled man to know of the existence of other wills-to-live. So the conflict can have a sort of resolution, reaching down to the smallest life: 'If I rescue an insect from a pool of water, then life has given itself for life, and the self-contradiction of the will-to-live has been removed' (p. 257).

# Appendix 4
## Man and animal: the biblical tradition

Among the influences on our attitudes to the relation of man and animals must be included the biblical account. Although today not many take the account of man and animals in the Bible literally, yet the Judaeo-Christian tradition is often referred to by animal moralists (e.g. Regan and Singer 1989), and biblical language and images pervade our culture. 'Dominion' is the word most frequently cited; yet in fact that is not quite the simple picture that emerges if one turns to the actual texts.[1]

What is believed to be the earlier creation story (Gen. 2: 5–25) describes how man is specially created from the dust, as a mist rises on a hitherto unwatered earth. The garden of Eden is made for him; and when the animals are created, their 'naming' is given to him. Then, as there was no suitable companion for him among the animals, no 'help meet for him', woman was created, from his own substance. It is in the later account (Gen. 1: 26–30) that, after the rest of creation, God creates man (and hence woman too) 'in his own image', and gives him 'dominion' over every living thing. Every herb and every tree is also given to him, and to the animals, for 'meat'. After the Fall, before casting them forth, God makes for Adam and his wife coats of skins, and clothes them (Gen. 3: 21). Later, their children come to make offerings to God. It is the offering by Abel the keeper of sheep, bringing the 'firstlings of his flock and of the fat thereof', that the Lord prefers over Cain's 'fruit of the ground', provoking Cain's wrath and the first murder.

Then comes the flood, with Noah's binary collection of all the animals into the Ark. When it is over, God delivers into Noah's hands the beasts and fowls and fish. 'Fear and dread of man' shall be upon them, and every moving thing shall be food except 'flesh ... with the blood thereof' (one of the elements in the

[1] All biblical quotations are from the Revised Version (Oxford University Press, Oxford, 1923).

vegetarian tradition). He establishes a 'covenant', marked by the rainbow, that the waters shall never become a destroying flood again. Yet it is a covenant not only with the man Noah, but also with every living creature (Gen. 9: 2–12).

This primary story is clear enough (apart from some ambiguity over vegetarianism). God has dominion over man and woman; and in turn man and woman are distinct from the other living things, entitled to name them, and to rule them and to use them. Yet it must be noticed that the other living things have significance enough to share God's covenant with Noah.

That there is more than mere dominion and subservience between man and animals pervades the Old Testament. Thus the tribes are given animal nick-names: Judah the lion's whelp, Issachar a strong ass, and Naphthali 'the hind let loose' (Gen. 49). The ox and the ass, as well as man, are to rest on the seventh day (Exod. 23: 12). One is to help a fallen ass, even if it is that of one's enemy (Exod. 23: 5). It was through an ass that Balaam was taught not to pretend things to be other than they are (Num. 22: 21–35). 'A righteous man regardeth the life of his beast, but the tender mercies of the wicked are cruel' (Prov. 12: 10). The ox treading the corn is not to be deprived of his reward by being muzzled (Deut. 25: 4). The vividness and affection of the imagery reflects the outlook: the stork, the turtledove, the swallow, the crane (Jer. 8: 7), Behemoth and the Leviathan with whom God plays (Job 40–1; Ps. 104), ostrich, horse, wild asses, hawk, eagle (Job 39), the exemplary ant, conies, locusts, lizards, and greyhound are all used for illustration or to point a moral (Prov. 30: 24–8), to cite a few examples. Outstanding is Isaiah's image of the holy mountain, with wolf, leopard, calf, lion, cow, bear, and ox—with a child in charge—where 'they shall not hurt or destroy in all my holy mountain' (Isa. 65: 25).

The New Testament is even more explicit on both the absolute and the relative value of man and animal. 'Are not five sparrows sold for two farthings? and not one of them is forgotten in the sight of God. But the very hairs of your head are all numbered. Fear not therefore, ye are of more value than many sparrows' (Luke 12: 6–7). 'Behold the birds of the heaven, that they sow not, neither do they reap nor gather into barns; and your heavenly father feedeth them. Are ye not of much more value than they?' (Matt. 6: 26). 'And he said to them, what man shall there be of you, that shall have one sheep, and if this fall into a pit on the sabbath day will not lay hold on it and lift it out? How much then is a man of more value than a sheep' (Matt. 12: 11–12). But even if both relative and absolute value are clear, it is an animal that, in the closing book of the New Testament, furnishes one of the greatest images, that of the Lamb who, by his quiet acceptance of harm, can uniquely break the recurring cycle of injury and revenge.

The New Testament brings out another theme, that of children. Jeremy Bentham in his famous footnote, not normally regarded as frivolous, found a full-grown horse or dog beyond comparison more rational, as well as more 'conversable', than a month-old baby. With that remark he opened up the current debate among some philosophers balancing the worth of 'superior' animals against the worth of the senile, the diseased, the handicapped, or the infantile human (see pp. 42–5). But in the New Testament it is the child who is of outstanding value. Christ especially welcomed and blessed children, for of such is the kingdom of God. The kingdom cannot be entered save as a little child. Their angels are always in God's presence. As for anyone who harms them, it were better for a millstone to be hanged about his neck and that he were drowned (Matt. 18: 6, 10 and 19: 13–14; Mark 9: 42 and 10: 13–16; Luke 17: 29). We are left to decide for ourselves what gives a child this moral centrality.

The underlying thoughts, perhaps, are: from the Old Testament, trusteeship, and from the New, the value of all creation, but especially of man, exemplified not by the learned or the powerful but by the child. While other religions and philosophies have generated other views of man and animal, these concepts still deserve their place when we think of that relationship today.

# Appendix 5
## Declaration of Helsinki (1964): Extract from 'Recommendations guiding doctors in clinical research'

### Introduction

It is the mission of the doctor to safeguard the health of the people. His knowledge and conscience are dedicated to the fulfilment of this mission.

The Declaration of Geneva of the World Health Medical Association binds the doctor with the words 'The health of my patient will be my first consideration'; and the International Code of Medical Ethics declares that 'Any act or advice which could weaken physical or mental resistance of a human being may be used only in his interest'.

Because it is essential that the results of laboratory experiments be applied to human beings to further scientific knowledge and to help suffering humanity, the World Medical Association has prepared the following recommendations as a guide to each doctor in clinical research. It must be stressed that the standards as drafted are only a guide to physicians all over the world. Doctors are not relieved from criminal, civil, and ethical responsibilities under the laws of their own countries.

In the field of clinical research a fundamental distinction must be recognized between clinical research in which the aim is essentially therapeutic for a patient, and the clinical research, the essential object of which is purely scientific and without therapeutic value to the person subjected to the research.

### Basic Principles

1. Clinical research must conform to the moral and scientific principles that justify medical research and should be based on laboratory and animal experiments or other scientifically established facts.

2. Clinical research should be conducted only by scientifically

qualified persons and under the supervision of a medically qualified medical man.

3. Clinical research cannot legitimately be carried out unless the importance of the objective is in proportion to the inherent risk to the subject.

4. Every clinical research project should be preceded by careful assessment of inherent risks in comparison to foreseeable benefits to the subject or to others.

5. Special caution should be exercised by the doctor in performing clinical research in which the personality of the subject is liable to be altered by drugs or experimental procedures.

# Appendix 6
## The Delaney Amendment (1958)

The Delaney Amendment to the Food Additives Amendment of 1958 comprises Section 409 (c) (3) (A) of the Federal Food, Drug, and Cosmetic Act. Its text reads as follows, beginning with 'Provided':[1]

(3) No such regulation shall issue if a fair evaluation before the Secretary—

(a) fails to establish that the proposed use of the food additive, under the conditions of use to be specified in the regulation, will be safe: *Provided*, That no additive shall be deemed to be safe if it is found to induce cancer when ingested by man or animal, or if it is found, after tests which are appropriate for the evaluation of the safety of food additives, to induce cancer in man or animal, except that this proviso shall not apply with respect to the use of a substance as an ingredient of feed for animals which are raised for food production, if the Secretary finds

(i) that, under the conditions of use and feeding specified in proposed labeling and reasonably certain to be followed in practice, such additive will not adversely affect the animals for which the feed is intended, and

(ii) that no residue of the additive will be found (by methods of examination prescribed or approved by the Secretary by regulations, which regulations shall not be subject to subsections (f) and (g) in any edible portion of such animal after slaughter or in any food yielded by or derived from the living animal.

[1] From Citizen's Commission on Science, Law and the Food Supply (1974), Rockefeller University, New York City, 'Report on Recent Symposia and Public Conferences, 23 April 1974'

# Notes

Full details of the works referred to here are given in the Bibliography (pp. 265–276).

## Chapter 1 (pp. 1–8)

1 Thomas 1984.
2 Maehle and Tröhler 1987.
3 Hall 1831:2–8.
4 May 1988.
5 Macaulay *c.*1876; Philanthropos 1883; Paget 1900; Smith *c.*1901; Coleridge 1906; Westacott 1949; French 1975; Moss 1961.
6 Walder 1983.
7 Ryder 1974:2.
8 Brock 1975.
9 CRAE 1977. This report on the LD50 is remarkable for the total confusion about the test, shown at the top of p. 6. Anyone taking the report seriously should first compare Trevan's S-shaped dose–response curve with the curve purporting to represent it, there displayed. The curve should be reversed, the ordinates interchanged and labelled and the ends of the curve made asymptotic to zero and 100% respectively. The report is useful, however, in listing CRAE's self-appointed structure at that date, which presumably approved the report's contents.

## Chapter 2 (pp. 9–21)

1 Orwell 1945.
2 Leeuwenhoek, in Dobell 1932.
3 Schweitzer 1933.
4 Eiseley 1959.
5 Darwin 1875:462.
6 Kavaliers *et al.* 1983.
7 Smith 1776.
8 Regan 1989:30.
9 French 1975.
10 Cruelty to Animals Act 1876.
11 French 1975.
12 Windsor 1988.

13 Andrews and Tansey 1981.
14 Ibid.
15 European Convention 1983.
16 Moss 1961.

Chapter 3 (pp. 22–29)

 1 Dale 1953.
 2 Pasteur 1854.
 3 Rothschild 1971.
 4 Dainton 1971.

Chapter 4 (pp. 30–54)

 1 Mountcastle 1968, ch. 63; Iggo *et al.* 1985.
 2 Protheroe 1991.
 3 Kavaliers *et al.* 1983, Kavaliers *et al.* 1984, Djoungoz *et al.* 1981.
 4 Lewis 1942.
 5 Royal Commission on Vivisection 1907–12; Macaulay *c.*1876; Westacott 1949.
 6 Burney, in Hemlow 1986.
 7 e.g. Coleman 1991. A remarkable case, where, among scores of allegations, almost the only adequately defined references were to the author's own numerous publications.
 8 *Oxford Times* 1979.
 9 Bentham 1789, trans. 1983.
10 Ibid., pp. 282–3.
11 Singer 1976.
12 Caplan 1983.
13 Frey 1983.
14 Regan 1989.
15 Regan 1983:78.
16 Ibid., p. 351.
17 Hull 1976.
18 e.g. in Langley 1989:12 (Midgley), 26 ff. (Regan); in Regan and Singer (1989): 83 (Singer), 227 (Frey).
19 Vanstone 1982:58–68.
20 Bryant 1991.
21 Vanstone 1982:58–68.
22 Ryder 1975.
23 Regan and Singer 1989:273; Langley 1989:15, 24.
24 Abbott 1884.
25 Linzey 1976:22.
26 Ritchie 1916.
27 Ibid., p. 81.

28 Caplan 1983.
29 Clark 1982:124.
30 Bentham 1789:205, note e2.
31 Rawls 1957.
32 See e.g. Brownlee 1971.
33 Schweitzer 1933, cited Appendix 3.
34 See Regan and Singer 1989:1–12; Baker 1975; Linzey 1976, Clark 1982.
35 See Greenpeace 1989–91.
36 I owe this point to Dr J. H. Botting.
37 Gladstone 1896.
38 See also Halsbury 1973, Dunstan 1979, Diamond 1981, Warnock 1986.

## Chapter 5 (pp. 55–107)

1 Kennedy 1968:272.
2 Garrison 1924; Majno 1975.
3 North 1983:82.
4 Paris 1822.
5 Hart 1946.
6 Office of Health Economics 1966.
7 Anderson 1977.
8 Office of Health Economics 1966.
9 *Lancet* 1985.
10 McKeown 1976.
11 Wilson and Miles 1955:1482–1551.
12 Hill 1962.
13 Wilson and Miles 1955:1955–81.
14 Crampton Smith 1971.
15 Keir 1950.
16 Paton 1976a.
17 Wilson and Miles 1955: 2165–75.
18 Medical Research Council 1977.
19 Billsborough 1983.
20 Pollock and Morris 1983.
21 *British Medical Journal* 1983.
22 Anderson and May 1982.
23 Rogers 1937.
24 Lehner *et al.* 1980.
25 Paton *et al.* 1978.
26 Burchenal and Krakoff 1956.
27 Shadwell 1911.
28 Doll and Peto 1981.

29 Ibid.

30 Doll 1983:93.

31 Learmonth 1954; Adams and Bell 1977.

32 Wiles and Devas 1954.

33 *Science* 1991.

34 Comroe and Dripps 1977; Lapage 1960; Barrett 1955.

35 *Lancet* 1991.

36 Brent 1965; Woodruff 1972; Calne 1984.

37 Beeson 1980a.

38 Osler 1909.

39 Browne 1658, ch. 5.

40 Weatherall 1982.

41 Geddes 1981.

42 *Anaesthesia News* 1991.

43 e.g. *British Pharmacopoeia 1982* and *British Pharmacopoeia (Veterinary) 1977*.

44 Herriot 1972, 1974.

45 Ewald and Gregg 1983.

46 Rogers 1937.

47 Jarrett 1990.

48 Goodwin 1980.

49 Shaper 1972.

50 Hull *et al.* 1983.

51 *British Medical Journal* 1975.

52 Lainson 1982.

53 Moulton 1907–12. The first quotation is from paragraph 12790, the second from 12737.

54 It is quite hard to recapture the past; but to read *Recent Advances in Chemotherapy* 1930 is a remarkable experience. It is sometimes claimed that no attempts at chemotherapy were being made before the days of sulfonamides and antibiotics. Here is an authoritative and fascinating 500-page record of decades of effort—with little hint of what was soon to come.

## Chapter 6 (pp. 108–121)

1 Comroe and Dripps 1977.

2 Doll 1983.

3 Quoted in Sharpe 1989:88.

4 Paton 1979b.

5 Osler 1909.

## Chapter 7 (pp. 122–129)

1 Lewis 1942, p. v.
2 Dawkins 1980.
3 Medawar 1972:30; see also 'On the use of animals in medical research', *Medical and Health Annual, Encyclopaedia Britannica* (Chicago 1983).

## Chapter 8 (pp. 130–161)

1 Littlewood 1965, Paras. 248–56 and General Finding 6.
2 Bawden and Brock 1982.
3 Smyth 1978.
4 Paton 1957.
5 European Medical Research Councils 1988.
6 Holmstedt and Liljestrand 1963.
7 Campbell *et al.* 1983.
8 Foster and Langley 1876.
9 Home Office 1983, Table 21.
10 Medical Research Council Annual Reports 1963–91.
11 Gowans 1974.
12 Paton 1979b.
13 I am indebted to Dr J. H. Botting for making copies of relevant sections of the *British Pharmacopoeia* for me.
14 Griffin 1991, Table 1, and Office of Health Economics 1983, Figure 5.
15 I am indebted to Dr D. B. Jeffreys of the Medicines Control Agency for these figures.
16 But see Jefferson 1955, Lepage 1960, Howard-Jones 1982.
17 Pappworth 1967; see also Elliott 1974.
18 Garrison 1924.
19 Holmstedt and Lijestrand 1963: Serturner's experiment is described on p. 74, Purkinje's on p. 87, and Christison's on p. 95.
20 Smith *et al.* 1947.
21 Head 1920:225–329; Trotter and Davies 1909.
22 Lewis 1927.
23 Lewis 1942.
24 Keele and Armstrong 1963.
25 Mountcastle 1968.
26 Torebjörk 1985.
27 Medical Research Council 1977. J. B. S. Haldane's work in diving has been well publicized, but many others have been similarly involved with little record.

## Chapter 9 (pp. 162–168)

1 Searle *et al.* 1989.
2 Spinks 1963.
3 Lester and Keoskowsky 1967.
4 Paton and Perry 1951.
5 Florey 1953.
6 Florey *et al.* 1949.
7 Botting 1991.

## Chapter 10 (pp. 169–194)

1 Home Office 1979.
2 Purchase 1990.
3 Holmstedt and Liljestrand 1963.
4 Trevan 1927 and Gaddum 1933.
5 British Toxicology Society 1984.
6 van den Heuvel *et al.* 1987.
7 van den Heuvel *et al.* 1990.
8 Uvarov 1985.
9 *Which?* 1991.
10 US House of Representatives 1952.
11 US Committee on Labor and Public Welfare 1976.
12 Draize *et al.* 1944.
13 Fielder *et al.* 1987.
14 Draize *et al.* 1944.
15 Barlow *et al.* 1991; Oliver 1990.
16 British Toxicology Society 1991.
17 Parsons *et al.* 1990; Peters and Piersma 1990.
18 Infante 1991.
19 Levine *et al.* 1991.
20 Ashby and Morrod 1991.
21 Department of Prices and Consumer Protection 1976.
22 Volans 1990.
23 *Science* 1987.
24 Starr 1969.
25 *Science* 1987.
26 Binns 1987.
27 Hutt 1985.

## Chapter 11 (pp. 195–224)

1 Balls 1988.
2 Green 1982; *Frame News* 1990; Medical Research Council 1991.
3 Pasteur 1854.

4 Langley (ed.) 1989; Midgley 1989.
5 *Conquest* 1977.
6 *Conquest* 1976.
7 Council of Europe 1983.
8 Bankowski and Howard-Jones 1984.
9 Sechzer 1983.
10 In Sperlinger 1981.

## Chapter 12 (pp. 225–230)

1 House of Lords 1980.
2 *European Convention* 1983.
3 Since the first edition of this book, the 1986 Act has incorporated all these points, as well as others then mentioned. The four general points now listed still seem worth stressing, however, since there may be further changes in the law.
4 Green 1982.
5 UFAW 1967.
6 Turner 1983; Sechzer 1983; Williams *et al.* 1983; *Toxicology in vitro, passim.*
7 Moulton 1907–12, para. 12704.
8 Russell and Burch 1959; *Frame News* nos. 24 and 25, and Greenpeace 1989–91 (revealing current policy).
9 Greenpeace 1989–91.
10 Darwin 1871.
11 Lovelock 1991.
12 Raup 1991.
13 Kaplan 1988.

# Bibliography

Abbott, E. A. (1884), *Flatland: a Romance of Many Dimensions, by the Author, a Square* (Seeley and Co., London).

Adams, I. W., and Bell, M. S. (1977), 'A comparative trial of polyglycolic acid and silk as suture materials for accidental wounds', *Lancet*, **ii**, 1216–17.

Altman, D. G. (1980), 'Statistics and ethics in medical research: VI Presentation of results', *British Medical Journal*, **281**, 1542–4.

American Association for the Advancement of Science (1990), *Resolution on Animal Experiment*; cited in *Frame News*, no. 26, pp. 8–9.

*Anaesthesia News* (1991), 'The vets are with us all the way', September issue.

Anderson, R. M., and May, R. M. (1982), 'Directly transmitted infectious diseases: control by vaccination', *Science*, **215**, 1053–60.

Anderson, T. (1977), 'The role of medicine', *Lancet*, **i**, 747.

Andrews, P. R. L., and Tansey, E. M. (1981), 'The effect of some anaesthetic agents on *Octopus vulgaris*', *Comparative Biochemistry and Physiology*, **70C**, 241–7.

Ashby, J., and Morrod, R. S. (1991), 'Detection of human carcinogens', *Nature*, **352**, 185.

Baker, J. A. (1975), 'Biblical attitudes to nature', in *Man and Nature*, ed. H. Montefiore (Collins, London), pp. 87–109.

Balls, E. M. (1988), 'Animal experiment—a necessary evil?', *Frame News*, **20**, 5, and 'Frame is not rich', ibid. 3.

Balls, M., and Clothier, R. H. (1991), 'Comments on the scientific validation and regulatory acceptance of *in vitro* toxicity tests', *Toxicology in vitro*, **5**, 535–8.

Bankowski, Z., and Howard-Jones, N. (1984), *Biomedical Research Involving Animals: Proposed International Guiding Principles*, (CIOMS, Geneva).

Barlow, A., Hirst, R., Pemberton, M. A., Hall, T. J., Oliver, G. J. A., and Botham, P. A. (1991), 'Refinement of an *in vitro* test for the identification of skin corrosive materials', *Toxicology Methods*, **1**, (2).

Barrett, N. R. (1955) (ed.), 'Surgery of the heart and thoracic blood vessels', *British Medical Bulletin*, **11**, 171–242.

Bawden, D., and Brock, Alison M. (1982), 'Chemical toxicology searching: a collaborative evaluation, comparing information resources and searching techniques', *Journal of Information Science*, **5**, 3–18.

Beeson, P. B. (1980a), 'Changes in medical therapy during the past half century', *Medicine*, **59**, 79–99.

Beeson, P. B. (1980b), 'How to foster the gain of knowledge about disease', *Perspectives in Biology and Medicine*, Winter 1980, Part 2.

Bentham, J. (1789), *An Introduction to the Principles of Morals and Legislation*, ed. J. H. Burns and H. L. A. Hart, 1983 (Methuen, London).

Billsborough, J. S. (1983), 'Whooping cough: the public health viewpoint', *Health Trends*, **15**, (3), 71–3.

Binns, T. B. (1987), 'Conference: therapeutic risks in perspective', *Lancet*, 25 July, p. 208.

Botham, P. (1990), 'Acute toxicity testing—scandal or science?', *Focus on Toxicology*, **6**, 3–5 (CTL, Alderley Park).

Botting, J. H. (1991), 'Penicillin: myth and fact', *R. D. S. Newsletter*, June 1991, pp. 8–9.

Brent, L. (1965), 'Transplantation of tissues and organs', *British Medical Bulletin*, **21**, 97–182.

British Association for the Advancement of Science (1982), *Experiments on Living Animals*, Report of a Symposium held in January 1982 (London).

*British Medical Journal* (1975), 'Immunisation in the two-thirds world' (editorial), v, 369.

*British Medical Journal* (1983), 'Views', i, 1288.

*British Pharmacopoeia 1982* (HMSO, London).

*British Pharmacopoeia (Veterinary) 1977* (HMSO, London).

British Toxicology Society (1984), 'A new approach to the classification of substances and preparations on the basis of their acute toxicity', report by Van den Heuvel, M., Clark, B., and Paton, W. D. M., *Human Toxicology*, **3**, 85–92.

British Toxicology Society (1991), 'Symposium on Immunotoxicity and Allergy (Abstracts)', *Human Toxicology*, **10**, 461–82.

Brock, Lord (1975), *Hansard*, 14 May, cols. 758–9; *The Times*, 8 May 1975, 23 October 1981, 24 February 1989; *Times Higher Educational Supplement*, 5 February 1982.

Browne, T. (1643), *Religio Medici*.

Browne, T. (1658), *Hydriotaphia, Urne Buriall, or, a Discourse of the Sepulchrall Urnes lately found in Norfolk*.

Brownlee, I. (1971) (ed.), *Basic Documents on Human Rights* (Clarendon Press, Oxford).

Bryant, P. E. (1991), 'Face to face with babies', *Nature*, **350**, 319.

Burchenal, J. H., and Krakoff, I. H. (1956), 'Newer agents in the treatment of leukemia', *Archives of Internal Medicine*, **98**, 567–73.

Calne, R. (1984), 'Can medicine advance without experiments on animals?', *Conquest*, no. 173, 1–9.

Campbell, W. C., Fisher, M. H., Stapley, E. O., Albers-Schöberg, G., and Jacob, T. A. (1983), 'Ivermectin: a potent new antiparasitic agent', *Science*, 221, 823–8.

Caplan, A. J. (1983), 'Beastly conduct: ethical issues in animal experimentation', *Annals of the New York Academy of Sciences*, 406, 159–69.

Card, W. I., and Mooney, G. H. (1977), 'What is the monetary value of a human life?', *British Medical Journal*, 2, 1627–9.

Clark, S. R. L. (1982), *The Nature of the Beast: Are Animals Moral?* (Oxford University Press, Oxford and New York).

Coleman, V. (1991), *Why Animal Experiments must Stop* (Merlin Press, London).

Coleridge, S. (1906), *Vivisection: a Heartless Science* (John Lane, The Bodley Head, London).

Comroe, J. H., and Dripps, R. D. (1977), *The Top Ten Clinical Advances in Cardiovascular-Pulmonary Medicine and Surgery 1945–1975* (US Government Printing Office, Washington DC).

*Conquest* (1976) no. 167, 'Some notes on experiments recently criticized'.

*Conquest* (1977) no. 168, 'Prosecutions under the 1876 Act'.

Council of Europe (1983), 'The use of live animals for experimental and industrial purposes', Ninth European Public Parliamentary Hearing, Strasbourg 8–9 December 1982.

CRAE (Committee for the Reform of Animal Experimentation) (1977), *The LD50 Test: Evidence for Submission to the Home Office Advisory Committee* (Private circulation).

Crampton Smith, A. (1971), 'Tetanus', in P. B. Beeson, and W. Macdermott, *Text-Book of Medicine (Cecil-Loeb)* (W. B. Saunders, Philadelphia), pp. 566–72.

Cruelty to Animals Act (1876), 39 and 40 Vict. ch. 77.

Dainton, Sir Frederick (1971), 'The future of the research council system', *A Framework for Government Research and Development*, Cmnd. 4814 (HMSO, London).

Dale, H. H. (1953), *Adventures in Physiology* (Pergamon Press, London).

Darwin, C. (1871), *The Descent of Man* (John Murray, London).

Darwin, C. (1875), *Insectivorous Plants* (John Murray, London).

Dawkins, Marion S. (1980), *Animal Suffering: the Science of Animal Welfare* (Chapman and Hall, London).

Department of Prices and Consumer Protection (1976), *Report by Consumer Safety Unit on Childhood Poisoning from Household Products* (Consumer Safety Unit, DPCP, London).

Diamond, C. (1981), 'Experimenting on animals: a problem in ethics', in D. Sperlinger (ed.), *Animals in Research*, pp. 337–62.

Djoungoz, M. B. A., *et al.* (1981), 'An opiate system in the goldfish retina', *Nature*, 292, 620–3.

Doll, R. (1983), 'Cancer control', The Lilly Lecture, *Symposium on Medical Management of Malignant Disease* (Royal College of Physicians, Edinburgh).

Doll, R., and Peto, R. (1981), *The Causes of Cancer* (Oxford University Press, Oxford and New York).

Draize, J. H., Woodard, E., and Calvery, H. O. (1944), 'Methods for the study of irritation and toxicity of substances applied topically to the skin', *Journal of Pharmacology and Experimental Therapeutics*, 82, 377–90.

Duffy, M. (1984), *Men and Beasts: An Animal Rights Handbook* (Paladin, London).

Dunstan, Canon G. R. (1979), 'A limited dominion', Paget Lecture, *Conquest*, no. 170, pp. 1–8.

Eiseley, Loren (1959), *Darwin's Century* (Gollancz, London).

Elliott, A. H. (1974), *Medical Experimentation* (Medical Research Council of New Zealand, PO Box 6063, Dunedin).

*European Convention for the Protection of Vertebrate Animals Used for Experimental and Other Scientific Purposes* (1983) (Council of Europe, Strasbourg).

European Medical Research Councils (1988), *Animal Experiments and Alternatives in Biomedical Research* (Quai Lezay-Marnist, F. 67,000, Strasbourg, France), European Science Foundation.

Ewald, B. H., and Gregg, D. A. (1983), 'Animal research for animals', *Annals of the New York Academy of Sciences*, 406, 48–58.

Fielder, R. J., Gaunt, I. F., Rhodes, C., Sullivan, F. M., and Swanston, D. W. (1987), 'A hierarchical approach to the assessment of dermal and ocular irritancy: a report by the British Toxicology Working Party on Irritancy', *Human Toxicology*, 6, 269–78.

Findlay, G. M., and Wenyon, C. M. (1930), *Recent Advances in Chemotherapy* (J. & A. Churchill, London).

Florey, H. W. (1953), 'The advance of chemotherapy by animal experiment', 21st Paget Lecture, *Conquest*, no. 41 (January), pp. 4–14.

Florey, H. W., Chain, E., Heatley, N. G., Jennings, M. A., Sanders, A. G., Abraham, E. P., and Florey, M. E. (1949), *Antibiotics* (Oxford University Press, Oxford), vol. ii, part 8.

Foster, M., and Langley, J. N. (1876), *A Course of Elementary Practical Physiology* (Macmillan, London).

*Frame News* (1990), 'Nigel Hamster Diary', 26, 5.

French, R. (1975), *Antivivisection and Medical Science in Victorian Society* (Princeton University Press, Princeton and London).

Frey, R. G. (1983), 'Vivisection, morals and medicine', *Journal of Medical Ethics*, 9, 94–7.

Gaddum, J. H. (1933), 'Methods of biological assay depending on a

Quantal Response', *Medical Research Council Special Report*, series no. 183 (HMSO, London).

Garrison, F. H. (1924), *History of Medicine* 3rd edition (Saunders, Philadelphia and London).

Geddes, A. M. (1981), 'Infection in Britain today', *Journal of the Royal College of Physicians, London*, **15**, 100.

Gladstone, W. E. (1896), *Studies Subsidiary to Butler's Works* (Clarendon Press, Oxford).

Goodwin, L. G. (1980), 'New drugs for old diseases', *Transactions of the Royal Society of Tropical Medicine and Hygiene*, **74**, 1–7.

Gowans, J. L. (1974), 'Alternative methods to animal experiment in medical research', Paget Lecture, *Conquest*, no. 165, pp. 2–6.

Green, C. J. (1982), *Animal Anaesthesia*, 2nd edition (Laboratory Animals Ltd., London).

Greenpeace (1989–91), *Something's happening* and *Against all odds*, campaign literature (Canonbury Villas, London N18BR).

Griffin, T. B. (1991), 'An economist's view of the pharmaceutical industry', *International Pharmacy Journal*, **5**, 206–8, Table 1.

Hall, Marshall (1831), *A Critical and Experimental Essay on the Circulation of the Blood* (Sherwood, Gilbert, and Piper, London); reprinted in 1847 in *Lancet*, **i**, 58–60, 135, 161.

Halsbury, Lord (1973), 'Ethics and the exploitation of animals', Paget Lecture, *Conquest*, no. 164, pp. 2–12.

Harris, H. (1970), *Cell Fusion* (Oxford University Press, Oxford).

Hart, P. D. (1946), 'Chemotherapy of tuberculosis', *British Medical Journal*, **ii**, 805.

Hart, P. D. (1988), 'The MRC and Tuberculosis Research', *M. R. C. News*, 75th Anniversary issue (Medical Research Council, London).

Harvard University, Office of Government and Community Affairs (1982), *The Animal Rights Movement in the United States: its Composition, Funding Sources, Goals, Strategies, and Potential Impact on Research*.

Head, H. (1920), *Studies in Neurology* (Oxford University Press, London), vol. i.

Hemlow, J. (1986), *Fanny Burney: Selected Letters and Journals* (Clarendon Press, Oxford).

Herriot, J. (1972), *It Shouldn't Happen to a Vet* (Michael Joseph, London).

Herriot, J. (1974), *Vet in Harness* (Michael Joseph, London).

Hill, A. Bradford (1962), *Statistical methods in clinical and preventive medicine* (E. & S. Livingstone, Edinburgh), reprint of 1952 article. 'Chemotherapy of pulmonary tuberculosis in young adults'.

Holmstedt, B., and Liljestrand, G. (1963), *Readings in Pharmacology* (Pergamon Press, Oxford and London).

Home Office (1979), *Report on the LD50 Test*, by the Advisory Committee on the Administration of the Cruelty to Animals Act 1876 (London).

House of Lords Select Committee on the Laboratory Animals Protection Bill (1980), vol. ii, *Minutes of Evidence* (HMSO, London).

Howard-Jones, N. (1982), 'Human experimentation in historical and ethical perspectives', *Social Science and Medicine*, **16**, 1429–48.

Hull, D. L. (1976), 'The rights of animals', *Science*, **192**, 679–80; review of *Animal Liberation* by P. Singer.

Hull, H. H., Williams, P. J., and Oldfield, F. (1983), 'Measles mortality and vaccine efficacy in rural West Africa', *Lancet*, **i**, 972.

Hutt, P. B. (1985), *Use of Quantitative Risk Assessment in Regulatory Decision-Making under Federal and Safety Statutes*, Banbury Report 19: Risk Quantitation and Regulation Policy (Cold Spring Harbour Laboratory, USA).

Iggo, A., Iversen, L. L., and Cervero, F. (1985), 'Nociception and pain', *Philosophical Transactions of the Royal Society of London*, **B308**, 217–431.

Infante, P. F. (1991), 'Prevention versus chemophobia: a defence of rodent carcinogenicity tests', *Lancet*, **337**, 538–54.

IVS (Index of Veterinary Specialities) (1991), *An Index of Ethical Preparations for the Veterinary Profession* 31/5 (A. E. Morgan Publications, Epsom).

Jarrett, W. F. H. (1990), 'Prospects of vaccines for AIDS and cervical cancer', *Conquest*, no. 179, 1–6.

Jefferson, Sir Geoffrey (1955), 'Man as an experimental animal', Paget Lecture, *Conquest*, no. 141, pp. 2–11.

Kaplan, J. (1988), 'The use of animals in research', *Science*, **242**, 839–40.

Kavaliers, M., Hirst, M., Teskey, G. C. (1983), 'A functional role for an opiate system in snail thermal behaviour', *Science*, **220**, 99–101.

Kavaliers, M., Hirst, M., Teskey, G. C. (1984), 'Opioid-induced feeding in the slug *Limax maximus*', *Physiology and Behaviour*, **33**, 765–7.

Keele, C. A., and Armstrong, Desirée (1963), *Substances Producing Pain and Itch*, Physiological Society Monograph no. 12 (Edward Arnold, London).

Keir, R. Macd, S. (1950), 'Case of tetanus treated with decamethonium iodide', *British Medical Journal*, **2**, 984–9.

Kennedy, M. (1968), *Portrait of Elgar* (Oxford University Press, Oxford).

Lainson, R. (1982), 'Leishmanial parasites of mammals in relation to human disease', *Symposium of the Zoological Society, London*, **50**, 137–79.

*Lancet* (1985), 'Decline in rheumatic fever', (editorial), 21 September, p. 647.

*Lancet* (1991), 'Cardiac myoplasty with the latissimus dorsi muscle', (editorial), **337**, 1383.

Langley, G. (1989), 'Plea for a Sensitive Science', in Langley (ed.), *Animal Experimentation: the Consensus Changes* (Macmillan Press, London). pp. 193–218.

Lapage, G. (1960), *Achievement. Some Contributions of Animal Experiment to the Conquest of Disease* (W. Heffer and Sons Ltd., Cambridge).

Learmonth, Sir James (1954), 'The surgeon's debt to animal experiment', *Conquest*, no. 137, pp. 3–10.

Leeuwenhoek, A. van, in Dobell, C. (1932), *Antony van Leeuwenhoek and his Little Animals* (Staples Prees, London).

Lehner, T., Russell, M. W., and Caldwell, J. (1980), 'Immunisation with a purified protein from *Streptococcus mutans* against dental caries in Rhesus monkeys', *Lancet*, i, 995–6.

Lester, D., and Keoskowsky, W. Z. (1967), 'Alcohol metabolism in the horse', *Life Science*, **6**, 2313–19.

Levine, A. J., Momaud, J., and Findlay, C. A. (1991), 'The p53 tumour suppressor gene', *Nature*, **351**, 453–6.

Lewis, T. (1927), *The Blood Vessels of the Human Skin and their Responses* (Shaw and Sons, London).

Lewis, T. (1942), *Pain* (Macmillan, New York).

Linzey, A. (1976), *Animal Rights* (SCM Press, London).

Littlewood, Sir Sydney (1965), *Report of the Departmental Committee on Experiments on Animals*, Cmnd. 2641 (HMSO, London).

Lovelock, J. (1991), *Gaia: the Practical Science of Planetary Medicine* (Gaia Books Ltd., London and Gloucester).

Macaulay, J. (no date given, *c.*1876), *Plea for Mercy to Animals* (Religious Tract Society, London).

McKeown, T. (1976), 'The role of medicine', Rock Carling Lecture (Nuffield Hospitals Provincial Trust, London).

Maehle, A.-H., and Tröhler, U. (1987), 'Animal experimentation from antiquity to the end of the 18th century: attitudes and arguments', in N. A. Rupke (ed.), *Vivisection in Historical Perspective*, pp. 14–47.

Majno, G. M. (1975), *The Healing Hand: Man and Wound in the Ancient World* (Harvard University Press, Cambridge, Mass.).

May, R. M. (1988), 'Control of feline delinquency', *Nature*, **332**, 392–3.

Medawar, P. B. (1972), *The Hope of Progress* (Methuen, London).

Medical Research Council (1947), 'Medical research in war', *Report of the Medical Research Council for the years 1939–1945*, Cmnd. 7335 (HMSO, London).

Medical Research Council (1963–91), *Annual Reports* (HMSO, London).

Medical Research Council (1977), 'Clinical trial of live measles vaccine given alone and live vaccine preceded by killed vaccine', Report by Measles Subcommittee, *Lancet*, ii, 571–5.

Medical Research Council (1991), *Report of MRC inquiry into the Operation of the Animals (Scientific Procedures) Act 1986 and the Administrative Procedures within the MRC in relation to Experiments carried out by Professor W. S. Feldberg* (MRC, London).

Midgley, M. (1989), 'Are you an animal?', in G. Langley (ed.), *Animal Experimentation*, pp. 1–18.

Morris, R. (1746), A Reasonable Plea for the Animal Creation: Being a Reply to a Late Pamphlet Intituled 'A Dissertation on the Voluntary Eating of Blood etc.' (London).

Morton, L. T. (1983), *A Medical Bibliography (Garrison and Morton)*, 4th edition (Gower Publishing, Aldershot).

Moss, A. W. (1961), *Valiant Crusade: The History of the R.S.P.C.A.* (Cassell, London).

Moulton, Lord Justice, in *Royal Commission on Vivisection* (1907–1912), paras. 12691–818.

Mountcastle, V. B. (1968), *Medical Physiology* (C. V. Mosby Company, St Louis).

New York Academy of Sciences (1988), *Interdisciplinary Principles and Guidelines for the Use of Animals in Research, Testing and Education*, Report by *ad hoc* Committee on Animal Research, 2 East 63rd St, New York.

North, R. (1983), *The Animals Report* (Penguin, Harmondsworth).

Office of Health Economics (1966), *Disorders which Shorten Life*, Report no. 21 (Office of Health Economics, London).

Office of Health Economics (1983), *Pharmaceutical Innovation*, Paper no. 74 (Whitehall, London).

Oliver, G. J. A. (1990), 'The evaluation of cutaneous toxicity: past and future', in *Skin Pharmacology and Toxicology*, ed. C. L. Galli, C. N. Hensby, and M. Marinovich (Plenum, New York).

Orwell, G. (1945), *Animal Farm* (Secker and Warburg, London).

Osler, W. (1909), 'The treatment of disease', *British Medical Journal*, ii, 185–9.

*Oxford Times* (1979), 'Reply on vivisection', 19 October.

Paget, S. (1900), *Experiments on Animals*, with an Introduction by Lord Lister (T. Fisher Unwin, London).

Pappworth, M. H. (1967), *Human Guinea Pigs: Experimentation on Man* (Routledge and Kegan Paul, London).

Paris, J. A. (1822), *Pharmacologia*, 5th edition (W. Phillips, London).

Parsons, J. F., Rockley, J., and Richold, M. (1990), '*In vitro* micromass teratogen test: interpretation of a blind trial of 25 compounds using three separate criteria', *Toxicology in vitro*, 4, 609–11.

Pasteur, L. (1854), 'Address at the opening of the Faculty of Letters (Douai) and of Sciences (Lille), 1854', in *Oeuvres de Pasteur*, ed. P. Valéry-Radot (Masson et Cie, Paris), vol. 7, pp. 129–32.

## 274  *Bibliography*

Paton, W. D. M. (1979a), 'Animal experiment and medical research: a study in evolution', Paget Lecture, *Conquest*, no. 169, pp. 1–14.

Paton, W. D. M. (1979b), 'The evolution of therapeutics: Osler's therapeutic nihilism and the changing pharmacopoeia', Osler Oration Lecture, *Journal of the Royal College of Physicians*, 13, 74–83.

Paton, W. D. M., and Perry, W. L. M. (1951), 'The pharmacology of the toxiferines', *British Journal of Pharmacology*, 6, 299–310.

Paton, W. D. M., Zaimis, Eleanor, Black, J. W., and Green, A. F. (1978), 'High blood pressure: the evolution of drug treatment: British contribution', in *Highlights of British Science* (The Royal Society, London).

Peters, P. W. J., and Piersma, A. H. (1990), '*In vitro* embryotoxicity and teratogenicity studies', *Toxicology in vitro*, 4, 570–6.

Philanthropos (1883), *Physiological Cruelty: or Fact v. Fancy* (Tinsley and Co., London).

Pollard, R. (1983), 'Whooping cough in Fiji', *Lancet*, i, 1381.

Pollock, T. M., and Morris, Jean (1983), 'A 7-year survey of disorders attributed to vaccination in North West Thames Region', *Lancet*, i, 753–7.

Protheroe, S. M. (1991), 'Congenital insensitivity to pain', *Journal of the Royal Society of Medicine*, 84, 558–9.

Purchase, I. F. R. (1990), 'Strategic considerations in industry's use of *in vitro* toxicology', *Toxicology in vitro*, 4, 667–74.

Raup, D. (1991), 'Extinction: bad genes or bad luck?', *New Scientist*, 14 September, pp. 46–9.

Rawls, J. (1957), 'Justice as fairness', *Journal of Philosophy*, 55, 653–62.

Regan, T. (1983), *The Case for Animal Rights* (University of California Press, Berkeley and Los Angeles), pp. 78, 351.

Regan, T. (1989), 'Ill-gotten gains', in G. Langley (ed.), *Animal Experimentation*, pp. 19–41; esp. pp. 30, 38–9.

Regan, T., and Singer, P. (1989) (eds.), *Animal Rights and Human Obligations*, 2nd edition (Prentice-Hall, Englewood Cliffs, N. J.), pp. 1–3, 6–12.

Research Defence Society (1991), 'Laboratory animal welfare: how Britain compares with other EEC countries', Newsletter, October, p. 13.

Ritchie, D. G. (1916), *Natural Rights*, 3rd edition (Allen and Unwin, London).

Rogers, Sir Leonard (1937), *The Truth about Vivisection* (Churchill, London).

Rothschild, Lord (1971), 'The organization and management of government research and development', in *A Framework for*

*Government Research and Development*, Cmnd. 4814 (HMSO, London).

Royal Commission on Vivisection, *Reports I–VI* and *Final Report* (1907–12) (HMSO, London).

Royal Society (1983), Symposium on *The ethics of experimentation on living animals*, July 1983 (London).

Rupke, N. A. (1987), (ed.), *Vivisection in Historical Perspective* (Croom Helm, London).

Russell, W. M. S., and Burch, R. L. (1959), *The Principles of Human Experimental Technique* (Methuen, London).

Ryder, R. D. (1974), *Scientific Cruelty for Commercial Profit* (Scottish Society for the Prevention of Vivisection).

Ryder, R. D. (1975), *Victims of Science* (Davis-Poynter, London).

Sather, H., Miller, D., Nesbit, M., Heyn, Ruth, and Hammond, D. (1981), 'Differences in prognosis for boys and girls with acute lymphoblastic leukemia', *Lancet*, i, 739.

Schweitzer, A. (1922), *On the Edge of the Primeval Forest*, trans. C. T. Campion (A. & C. Black, London).

Schweitzer, A. (1923), *Civilization and Ethics*, Dale Memorial Lectures, Part 2, trans. J. Naish (A. & C. Black, London).

Schweitzer, A. (1924), *Memoirs of Childhood and Youth* (George Allen and Unwin, London).

Schweitzer, A. (1933), *My Life and Thought: an Autobiography*, trans. C. T. Campion (George Allen and Unwin, London).

*Science* (1987), Symposium on Risk, **236**, 267–300; **242**, 44–9.

*Science* (1991), 'Implant Materials', **252**, 1057.

Searle, A. G., Peter, J., Lyon, M. F., Hall, J. G., Evans, E. P., Edwards, J. H., and Buckle, V. J. (1989), 'Chromosome maps of man and mouse IV', *Annals of Human Genetics*, **53**, 89–140.

Sechzer, J. (1981), 'Historical issues concerning animal experimentation in the United States', *Social Science and Medicine*, **15F**, 13–17.

Sechzer, Jeri A. (1983) (ed.), 'The role of animals in biomedical research', *Annals of the New York Academy of Sciences*, **406**, 1–229.

Shadwell, A. (1911), 'Cancer', in *Encyclopaedia Britannica*, 11th edition (Cambridge University Press, Cambridge).

Shaper, A. G. (1972), 'Cardiovascular disease in the tropics. I. Rheumatic heart', *British Medical Journal*, iii, 683–6.

Sharpe, R. (1989), 'Animal experiments—failed technology', in G. Langley (ed.), *Animal Experimentation*, pp. 88–117.

Singer, P. (1976), *Animal Liberation* (Jonathan Cape, London).

Smith, Adam (1776), *An Enquiry into the Nature and Causes of the Wealth of Nations*, 1890 edn. (Routledge, London).

Smith, S. (no date given, c.1901), *Scientific Research: a View from Within* (Elliott Stock, London).

Smith, S. M., Brown, H. O., Toman, J. E. P., and Goodman, L. S. (1947), 'Lack of cerebral effects of d-tubocurarine', *Anaesthesiology*, **8**, 1–14.

Smyth, D. H. (1978), *Alternatives to Animal Experiment* (Scolar Press, London).

Sperlinger, D. (1981) (ed.), *Animals in Research: New Perspectives in Animal Experimentation* (John Wiley & Sons, Chichester & New York).

Spinks, A. (1963), 'Justification of clinical trial of new drugs', *Proceedings of the Second International Pharmacological Meeting* (Prague), **8**, 7–19.

Starr, C. (1969), 'Social benefits versus technological risk: what is our society willing to pay for safety?', *Science*, **165**, 1232–8.

Thomas, K. (1984), *Man and the Natural World* (Penguin, Harmondsworth).

Torebjörk, E. (1985), 'Nociceptor activation and pain', in Iggo *et al.*, 'Nociception and pain', pp. 227–34.

Trevan, J. W. (1927), 'The error of determination of toxicity', *Proceedings of the Royal Society*, B, **101**, 483–514.

Trotter, W., and Davies, H. M. (1909), 'Experimental studies on the innervation of the skin', *Journal of Physiology*, **38**, 134.

Turner, P. (1983) (ed.), *Animals in Scientific Research: an Effective Substitute for Man?* (Macmillan, London).

UFAW Handbook (1967), *The Care and Management of Laboratory Animals*, 3rd edition (Livingstone, Edinburgh and London).

US Committee on Labor and Public Welfare, Cosmetic Safety Amendments (1976), Hearing before the Subcommittee on Health (US Government Printing Office, Washington DC).

US House of Representatives (1952), Hearings by Select Committee to Investigate the Use of Chemicals in Foods and Cosmetics (US Government Printing Office, Washington DC).

Uvarov, O. (1985), 'Research with animals: requirement, responsibility, welfare', *Laboratory Animals*, **19**, 51–75.

Vanstone, W. H. (1982), *The Stature of Waiting* (Darton, Longman and Todd, London).

Volans, G. N. (1990), 'Chemical poisoning—the role of the National Poisons Boards', *Focus on Toxicology*, **6**, 10–12 (CTL, Alderley Park).

Van den Heuvel, M., Clark, D., Fielder, R., Pelling, D., Tomlinson, N., and Walker, A. (1990), 'The international validation of a fixed-dose procedure as an alternative to the classical LD50 test', *Food and Chemical Toxicology*, **28**, 469–82.

Van den Heuvel, M., Dayan, A. D., and Shikkaker, R. D. (1987), 'Evaluation of the BTS approach to the testing of substances and preparations for their acute toxicity', *Human Toxicology*, **6**, 279–91.

Van den Heuvel, M., and Fielder, R. J. (1990), 'Acceptance of *in vitro* testing by regulatory authorities', *Toxicology in vitro*, **4**, 675–9.

Walder, A. (1983), quoted in *The Times*, 22 April.

Warnock, M. (1986), 'Law and the pursuit of knowledge', *Conquest*, no. 175, pp. 1–7.

Weatherall, Josephine S. (1982), 'A Review of some effects of recent medical practices in reducing the numbers of children born with congenital abnormalities', *Health Trends*, 14(4), 85–8.

Westacoot, E. (1949), *A Century of Vivisection and Antivivisection* (C. W. Daniel Company Ltd., Ashington, England).

*Which?* (1991), 'Animals: your rights', p. 19, January; 'Chemical hazards', p. 210, April; 'Suntan lotions', p. 330, June; 'Washing-up liquids', p. 366, and 'Insect repellents', p. 398, July; 'Cellulite treatments', p. 456, August; 'Contact lenses and solutions', p. 497, September.

Wiles, P., and Devas, M. B. (1954), 'The halt and the maimed', *Conquest*, no. 138, pp. 2–6.

Williams, G. M., Dunkel, V. C., and Ray, V. A. (1983) (eds.), 'Cellular systems for toxicity testing', *Annals of the New York Academy of Sciences*, **407**, 1–484.

Wilson, G. S., and Miles, A. A. (1955), *Principles of Bacteriology and Immunity*, 4th edition (Edward Arnold, London), pp. 1482–1551, 1955–81, 2165–75.

Windsor, C. (1988), 'Recalling memories', *Physics Bulletin*, **39**, 16–18.

Woodruff, Sir Michael (1972), 'The contribution of animal experiments to the surgery of replacement', *Conquest*, no. 163, pp. 3–7.

# Index